Library of
Davidson College

SEVENTY-FIVE YEARS OF PROGRESS

Richard K. Meade,
Auditor, 1908-1909

John C. Olsen,
President, 1931
Secretary, 1909-1926

William H. Walker

Charles F. McKenna,
President, 1910

Arthur D. Little,
President, 1919

William M. Booth,
Treasurer, 1909-1910

The Committee of Six

75 YEARS OF PROGRESS

–a history of the
American Institute of Chemical Engineers

1908–1983

by
Terry S. Reynolds
Department of General Engineering
University of Wisconsin–Madison

Published by
AMERICAN INSTITUTE OF CHEMICAL ENGINEERS
345 East 47th St., New York, New York 10017

Edited by
J. Charles Forman, *Executive Director*
Larry Resen, *Publications Director*

Copyright, 1983

Library of Congress Cataloging in Publication Data

Reynolds, Terry S.
 75 Years of Progress.

 Includes bibliographical references.
 1. American Institute of Chemical Engineers—History.
I. Title. II. Title: Seventy-five Years of Progress.
TP1.A636R49 1983   660.2'06'073   83-11875
ISBN 0-8169-0231-3

Foreword

To some extent this work can be regarded as a companion volume to F. J. Van Antwerpen and Sylvia Fourdrinier's *High Lights: The First Fifty Years of the American Institute of Chemical Engineers* (1908–1958). Although I also deal with the years 1908 to 1958 in this book, my approach to this period is different, and those interested in more details on certain topics (for example, awards) may find the earlier publication an interesting additional reference.

This work is based on three principal sources: (1) the work mentioned in the preceding paragraph, (2) the published records of the Institute, and (3) the combined *Minutes* of the Institute's Council and Executive Committee. In general, for the period 1908 to 1955 I have relied more heavily on the first two sources than on the third. There are good reasons for this. Prior to 1955 the *Minutes* are sketchy and not very informative. Much of the Institute's business during its first four decades, moreover, was carried out during national meetings of the organization and was reported in the "Proceedings" printed in the AIChE *Bulletin*. Only after 1955 do the *Minutes* become more informative than the published record.

For providing assistance during my research in AIChE headquarters in New York, I would like to acknowledge Harold Abramson, Director of Educational Services; Rose Burgos, Editorial Assistant; Bob Guertin of the Controller's Office; Trudy Hutter of Admissions; Bob Kenney, Office Manager; Gloria Lambson, Director of Member Activities and Services; Valeria Pastyik, Assistant to the Executive Director; Raymond Rivera of the Mail Room; and Karen Simpson, Editorial Assistant. For reading and providing comments on my initial manuscript I thank J. Charles Forman, Executive Director of AIChE; Richard R. Hughes, 1982 AIChE President; Robert H. Marshall, 1983 AIChE President; Larry Resen, Director of Publications; Klaus Timmerhaus, 1975 AIChE President; and A. S. West, 1976 AIChE President.

I owe special thanks to Professor Jean-Claude Guédon of the University of Montreal, who provided me with an English-language manuscript of his article on the emergence of unit operations; to W. Robert Marshall, Jr., 1963 AIChE President and my former Dean, who not only read and commented on my manuscript but took time on a number of occasions to answer questions about the AIChE and its operations; to Franklin J. VanAntwerpen, former Executive Secretary/Director of the American Institute of Chemical Engineers; and to Linda Gail Rainwater Reynolds, my wife and a partner in all of my work. Regarding the record-keeping, the AIChE Council and Executive Committee *Minutes* before 1955, when Van Antwerpen became Executive Secretary, are so brief that it is usually impossible to determine the motivations or stimuli behind Institute actions. The *Minutes* left by Van Antwerpen and his successor, J. Charles Forman, on the other hand, are complete. They usually indicate what prompted actions, summarize the course of Council and Executive Committee debates, and indicate reservations and disagreements when reservations and disagreements existed. Moreover, all important committee reports, as well as other relevant documents, are appended to the *Minutes*. These more detailed records made my task immeasurably easier and my work much more complete and accurate than it otherwise would have been.

Terry S. Reynolds

Table of Contents

FOREWORD	i
CHAPTER 1	
The Struggle for Legitimacy, 1908–c1930	1
Notes	20
CHAPTER 2	
Coming of Age, c1930–c1958	26
Notes	48
CHAPTER 3	
Expanding Programs, c1958–c1970	53
Notes	84
CHAPTER 4	
Maintaining Balance, c1970–c1983	90
Notes	121
Postscript	128
Notes	129
CHAPTER 5	
Anecdotal Reminiscences	130
CHAPTER 6	
A Look to the Future	176
APPENDIX	179

LETTER SENT BY COMMITTEE OF SIX

New York, September 3, 1907

Dear Sir:

At Atlantic City, June 21, 1907, a number of chemists, mostly members of the American Society for Testing Materials, met and discussed the advisability of forming a society to be composed of trained chemists who are designing, constructing, managing or controlling establishments involving the application of chemical principals to the treatment of raw materials or doing work having an intimate connection with such operations; such a society to have a high standard of admission somewhat similar to the requirements of membership for the American Society of Civil Engineers. (10 years of practical experience, 5 of which being in a responsible position. Age limit for Junior Members 20 years, for Senior Members 30 years.)

The general opinion of those present at this meeting was that the formation of such a society would tend to raise the standing of chemists among manufacturers and that many other advantages would accrue among which the following were mentioned by various speakers:—

    a. Raising the standard of education of chemists.
    b. Raising the ethical standard of the profession.
    c. Encouraging bold and original work by members.

Quarterly national meetings were suggested to avoid the evils of localized control.

Some doubt was expressed as to whether the time had come for the formation of such a society because of possible detriment to the present very excellent chemical societies and also as to whether there is a general desire among chemists for its organization.

A committee was appointed to present the matter to a large number of American chemists and obtain their opinion as to the advisability of the formation at this time, and the conditions of membership of such a society.

This committee, therefore, addresses you and asks your opinion with reference to the following:—

1. Do you consider the formation of such a society advisable?
2. In your opinion is an organization on such lines likely to afford any special and important advantages?
3. Do you believe these advantages will continue in the future?
4. Would you favor the use of the title American Society of Chemical Engineers?
5. Would you favor a high standard of admission?
6. Would you favor an age limit for membership, 30 for instance?
7. Would you favor Junior and Senior grades of membership?
8. Would you favor a membership test by education, career or achievement, any or all of these?

The committee will appreciate an early reply and a full statement of your views on the questions asked and any other pertinent suggestions which you may wish to offer.

    (Signed by)

    Dr. Charles F. McKenna, Chairman
    221 Pearl St., New York, N. Y.

    William M. Booth, Secretary
    Syracuse, New York

    Prof. John C. Olsen
    Polytechnic Institute, B'klyn, N. Y.

    Richard K. Meade
    Nazareth, Pennsylvania

    Prof. William H. Walker
    Inst. of Technology, Boston, Mass.

    Arthur D. Little
    93 Broad St., Boston, Mass.

CHAPTER **1**

The Struggle for Legitimacy, 1908–c1930

BACKGROUND

The early history of the American Institute of Chemical Engineers was dominated by two problems—establishing exactly what chemical engineering was and establishing the legitimacy of chemical engineering as an independent discipline. The Institute, in other words, was compelled to stake out its territory in the realm of knowledge and justify its claims to this territory in the face of other disciplines claiming the same territory.

These were rather unique problems for an engineering society. Most of the other major engineering societies had not faced them. Because civil, mining, and mechanical engineering had extensive roots stretching back for centuries before the emergence of their professional societies, neither the American Society of Civil Engineers (ASCE), nor the American Institute of Mining and Metallurgical Engineers (AIME), nor the American Society of Mechanical Engineers (ASME) had much difficulty in securing recognition for their disciplines.[1]

The American Institute of Chemical Engineers, however, faced serious problems. Chemical engineering did not have a long history as a *recognized* discipline.[2] In fact, at the turn of the century there were three rival traditions which were prepared to argue that chemical engineering did not exist as a separate discipline and lay claim to its territory (i.e., to chemical manufacturing). Two of these traditions—industrial chemistry and applied chemistry—regarded manufacturing chemistry as the domain of chemistry. The industrial chemist and applied chemist approached manufacturing chemistry in a different manner than the chemical engineer would in later years. The former, for instance, tended to focus on the production of a single chemical from beginning to end, devoting little attention to the common elements (unit operations) in the manufacture of different chemicals. The latter tended to break industrial processes down into a series of chemical reactions, paying little attention to the technical problems involved in bulk production. A third tradition—mechanical engineering—also claimed much of the field that would later become chemical engineering. Like the applied chemist and the industrial chemist, the mechanical engineer's view of manufacturing chemistry differed from the chemical engineer's. The mechanical engineer focused on the types of machinery needed to carry out bulk production of chemicals, but tended to slight the chemical processed involved.[3] Had any of these traditions proved dominant, chemical engineering either would have never been born, or would have remained only a small, unimportant specialist discipline, or would have emerged much later than it did. In Germany, for example, where the tradition of industrial chemistry predominated, chemical engineering (*verfahrenstechnik*) was very late (largely post 1950) in emerging as a distinct discipline.[4]

Those who viewed chemical manufacturing as the realm of industrial or applied chemistry provided the strongest opposition in the United States to the claims of chemical engineering to disciplinary stature. At a 1904 meeting of the American Chemical Society (ACS), for instance, Hugo Schweitzer, a chemist, declared: "I am absolutely against the introduction of chemical engineering in the education of chemists." He urged, instead, that chemical manufacturers take pure chemists from the universities and train them in whatever technical aspects of their plants they desired. M. T. Bogert, later a president of the ACS (1907–08), spoke in a similar vein. He argued that progress in "technical chemistry" was due largely to work in research laboratories by men with no engineering training, and he pleaded with employers for recognition of the work of these men and greater patience in their dealings with them.[5]

Industrial Influences

Two factors, however, worked against the claims being made on the field of chemical manufacturing by the chemists—the rapid growth of the American chemical industry and the American tradition of mass production. The American chemical industry was negligible before 1880. But between 1880 and 1914, encouraged and protected by the high tariff policies of the Republican administrations of the era, growth was rapid. Estimates of the value of the products of the American chemical industry vary due to the industry's diverse nature. One set of data, however, indicates that in 1879 the chemical industry (excluding drugs, medicines, etc.) employed only 11,064 wage earners and manufactured products valued at only $44 million, while in 1914 the industry employed almost 45,000 wage earners and manufactured products valued at $221.5 million. Haber, relying on different statistics, estimated the value of products for chemical and allied industries in the United States at $840 million for 1914.[6]

In any case, by the onset of the First World War the American chemical industry had become either the largest or the second largest in the world.[7] In some areas it was clearly first—superphosphates, sulfuric acid, sodium compounds, and carbides, for instance. And in a number of other areas it was using production processes well in advance of the older European chemical industry.[8] This rapid growth created an increasing need for a class of professionals with engineering as well as chemical skills.

The American tradition of mass production was even more critical to the emergence of chemical engineering as a profession than the quantitative growth of the American chemical industry. Influenced by the success of Germany in dyestuffs, chemical manufacturing in Europe was largely oriented toward the introduction of new products and to high quality, low volume "fine" chemicals. In America the long-standing preference for more efficient mass production carried over into the chemical industry. The American chemical industry emphasized the production in quantity of already existing chemical products. This emphasis meant that the design, construction, and operation of large pieces of mechanical equipment (needed to handle large volumes of chemicals) was *the* central feature of American chemical plant design,[9] and the American chemical industry by 1900 urgently needed personnel with a strong combination of chemical and engineering skills.

This combination was not supplied by industrial chemistry, applied chemistry, or even mechanical engineering, neither was such a combination in demand in European countries where production volumes were much lower.[10]

Personnel with appropriate skills for the rapidly expanding, high volume American chemical industry were not initially available. Dissatisfaction with the shortcomings of the industrial and applied chemists traditionally employed in American chemical manufacturing operations was rife in the early twentieth century. For instance, at the same American Chemical Society meeting in 1904 that had brought appeals to chemical manufacturers to continue to take pure chemists from the universities and to be patient with them and educate them in the technical aspects of particular industries, M. C. Whitaker, General Superintendent of the Welsbach Co. and a future Perkins Medalist (1923), complained that the chemist was "generally not a man who is capable of transmitting from a laboratory to a factory the ideas which he has developed" because he was not educated "in the engineering branches."[11] Another spokesman, the manufacturer J. B. F. Herreshoff, complained that students were trained to look on the great pure chemists as "supreme beings" and thus considered mechanical engineering "as beneath the dignity of the chemist." As a result they left college "not knowing anything of mechanical engineering and were totally unfit to take positions as works managers or wherever it became necessary to apply chemistry in a large way."[12] Herreshoff added:

> The greater the application of chemistry, the more important becomes the combining of mechanical engineering with chemical engineering. Our colleges should consider this matter more seriously than ever, and do their best to make the course in chemical engineering as complete and perfect in every way possible.[13]

There had been attempts, beginning as early as 1888 at the Massachusetts Institute of Technology, to train engineers for the growing chemical industry. But these "chemical engineering" programs had had limited success, primarily because they were a hodgepodge of courses in chemistry and various engineering disciplines—usually mechanical and electrical—with few courses designed specifically for chemical engineers. As a result what usually emerged was either chemists with a smattering of engineering or mechanical engineers with a smattering of chemistry. These were scarcely more satisfactory than applied chemists or industrial chemists.[14]

It was against this background that the American Institute of Chemical Engineers was founded.

FOUNDING

In October, 1905, Richard K. Meade, founder of *The Chemical Engineer*, a periodical just finishing its first year of publication, asked in an editorial: "Why not the American Society of Chemical Engineers?"[15] He argued that a chemical engineering discipline had been slowly emerging, pointing to several colleges that had begun offering courses titled "chemical engineering" (although they were, in content, primarily industrial chemistry). He estimated that there were at least 500 chemical engineers already in the United States. Yet many chemical plants were being designed by mechanical engineers. A professional society, he argued, would help convince manufacturers that chemical engineers or chemists with an engineering background should design and operate these plants and would promote the exchange of ideas to mutual advantage.

In March, 1907, Meade reprinted his 1905 editorial.[16] He then wrote letters to 50 chemists and chemical engineers seeking support for the idea of founding an organization for the emerging discipline. Encouraged by the response, he took the responsibility of calling a preliminary meeting to discuss the subject.[17]

This preliminary meeting was held on June 21, 1907, in Atlantic City, N.J., in conjunction with a meeting of the American Society for Testing Materials. Eleven were present, counting Meade. The group discussed the objectives of a new technical society and, on motion from John C. Olsen of Brooklyn Polytechnic Institute, approved the formation of a six-man committee to further consider the advisability of forming a society for chemical engineers, correspond with interested parties, and establish possible membership criteria.[18] This Committee of Six met three times in 1907 and sent out questionnaires to 600 American and Canadian chemists, querying them on the advisability of forming a new society and on possible membership requirements. Of the 187 replies received, 70% favored the formation of a new society, 87% favored high entrance requirements.[19] In spite of this support, the committee was reluctant to make a final decision on its own. At a December 7, 1907, meeting at the Chemists' Club in New York, it decided to ask 50 prominent chemists and chemical engineers, including some known to be opposed to the formation of a new professional group, to serve on a committee to consider the formation of a society.[20]

This committee met in New York on January 18, 1908, with 21 in attendance. Charles F. McKenna, chairman of the preliminary meeting in Atlantic City and of the Committee of Six and 1910 President, presided over the session. After making brief opening remarks, he turned the floor over to the meeting's secretary, John Olsen, who read comments from correspondents on the advisability of forming or not forming a new society. The meeting was then thrown open for discussion.[21]

M. T. Bogert of Columbia, President of the American Chemical Society, distributed ACS Committee reports outlining the actions his organization intended to take to meet the needs of chemists working in industry, including the publication of a new *Journal of Industrial and Engineering Chemistry (I&EC)* and the organization of an autonomous Division of Industrial Chemists and Chemical Engineers. He was not the only one to express opposition to the projected new society. A. C. Langmuir, Chief Chemist and Superintendent for Marx & Rawolle, argued against using the term "chemical engineer," feeling that it was too broad and misleading. He supported Bogert, arguing that an industrial section of the ACS was sufficient to cover the needs of manufacturing chemistry. Another spokesman declared: " 'Chemist' is a good enough term for me, and if the chemists stand up for their names and their interests as well as the engineers do, you do not need any society to speak for you. Stick to what you have. I do not think we have any need to form any other organization."

But others spoke of the need for an intermediate discipline between chemistry and mechanical engineering and the need for engineers specially qualified to design and operate large-scale chemical manufacturing establishments. Olsen, for example, argued that the proposed division of the American Chemical Society was too broad and could not accomplish the same goals as a society "composed of men who can be called strictly engineers."

There was sufficient opposition to the proposal for creating a new society that those at the meeting, instead of reaching a decision, voted to put the question to a vote of the full Committee of Fifty by mail. Bogert, for the ACS, sent a circular following the meeting to the 50, detailing his objections to the creation of a new organization. Proponents of the proposed new society, somewhat angered by this action, responded with a circular of their own.[22] Of the ballots returned, 22 favored organizing a society for chemical engineers, 7 opposed the idea, 7 were neutral.[23]

The original Committee of Six met on April 25, 1908, and elected two additional members, one of whom, S. P. Sadtler, was asked to begin planning an organizational meeting for the proposed society. On May 25 the enlarged committee met and formally issued invitations to more than 100, asking them to meet at the Engineers' Club in Philadelphia on June 22, 1908.[24]

Nineteen people attended the organizational meeting of the American Institute of Chemical Engineers on a hot and humid day in Philadelphia.[25] McKenna opened the morning session with a keynote address justifying the creation of a society for chemical engineers.[26] That afternoon the group approved a constitution and elected the first slate of officers. Samuel Sadtler was elected president. John Olsen, who had been active throughout the planning process, was elected secretary, a post he was to hold until 1927.[27]

The constitution adopted at the organizational meeting of the AIChE in Philadelphia laid down very high membership requirements, but made the Institute internally more democratic than most contemporary engineering societies. It provided for a president, three vice presidents, a secretary, a treasurer, an auditor, and nine directors. The primary administrative body was to be a Council, made up of these officers, with the exception of the auditor. Candidates were to be solicited by nominating ballots mailed to each member eight weeks prior to the winter meeting. After these ballots were returned, the secretary was required by the constitution to prepare an election ballot which included the names of all candidates appearing on at least ten nominating ballots.[28] These provisions insured open, contested elections, a tradition which was to continue to be a hallmark of the AIChE.

THE AIChE AND THE ACS

At the January, 1908, meeting of the Committee of Fifty there had been some confusion over just what chemical engineering was and considerable opposition to the creation of a society devoted to such a specialty. Thus it must have been clear to the founders of the Institute that the new organization would have to clarify what it meant by chemical engineering and justify its claims to this territory. At the same time, the infant organization would have to avoid, as far as possible, direct confrontation with older and stronger claimants to the same territory. Translated into specifics this meant (1) that the AIChE had to minimize direct conflicts with the American Chemical Society, a larger, older, and stronger organization which claimed the same domain and had opposed the AIChE's creation and (2) that the new organization, through the creation of appropriate educational and technical programs, had to define more clearly what chemical engineering was and justify its legitimacy as an independent discipline.

Probably the most pressing of these two problems, initially at least, was that of minimizing conflict with the ACS. It was established in 1876 and was by the early twentieth century an important and powerful organization. By 1908 it had 5000 members and claimed the entire realm of chemistry, both pure and applied.[29] Moreover, at the time it looked on other societies in its very broad field as interlopers who threatened its programs and splintered what should have been a unified discipline. Its leadership had been sensitized to the problem of splinter societies by the secession of electrochemists in 1902 to form the American Electrochemical Society and leather chemists in 1904 to form the American Leather Chemists Association.[30]

It is not surprising, then, that the creation of the American Institute of Chemical Engineers was not only opposed, but viewed with alarm by ACS leaders like Bogert, who, as noted, argued strongly against the formation of the Institute in early 1908.[31] The creation of the AIChE helped galvanize the ACS to take action to prevent further losses of chemists with industrial interests. On June 30, 1908, eight days after the formal founding of the American Institute of Chemical Engineers, the ACS created its first technical division—the Division of Industrial Chemistry and Chemical Engineering—a division which claimed the same area as AIChE. In December, 1908, and January, 1909, the ACS established several other divisions that potentially overlapped the domain being claimed by the new chemical engineering society, for example, the Division of Fertilizer Chemistry.[32] Thus, in the words of one of AIChE's charter members, the new organization faced "strong protest and rather bitter feelings in certain quarters" at its founding.[33]

The Institute's leadership, however, did not wish their new organization to be viewed as a rival of the broader and more firmly established chemists' society. Most of the AIChE's founders and early members had been trained as chemists and had acquired engineering interests and skills only through practice. Most early Institute members, including the entire Committee of Six, 42 of the Committee of Fifty, and the entire first slate of officers, were also ACS members.[34] They did not wish to compel others or be compelled to choose themselves between the two groups. Many probably recognized that direct rivalry between the two groups would absorb resources which the infant AIChE did not have and potentially cripple both the emerging profession of chemical engineering and the organization claiming to be its spokesman. The Institute's claim on the loyalty of even some of its founders was weak enough without putting additional strains on it. Two members of the Committee of Six—Arthur D. Little and William H. Walker—were not charter members of the new organization (both did join some years later). Little (1919 President), instead, accepted the post of Chairman of the ACS's Division of Industrial Chemistry and Chemical Engineering, and both Little and Walker served on the editorial board of ACS's *Journal of Industrial and Engineering Chemistry (I&EC)*.[35]

For several decades the AIChE's leaders attempted to avoid direct rivalry in three manners: (1) by adopting very restrictive membership requirements, (2) by emphasizing the Institute's role as complementary to ACS, and (3) by approaching possible problem areas with caution and discretion.

Perhaps the most effective of these methods was the adoption of more restrictive entrance requirements. The ACS admitted to its membership anyone "interested in the promotion of chemistry," including teachers at all levels, amateur chemists, industrial chemists, and chemical technicians.[36] The AIChE adopted much more restrictive membership requirements. To become an "Active" member one had to be at least 30 years old and proficient in "chemistry *and* in some branch of engineering as applied to chemical problems." In addition, Active members were required, at the time of election, to be actively engaged in work involving the application of chemical principles to practice. Further, members were required to have ten years of practical experience in chemical manufacturing if they held no academic degree, with five years in charge of manufacturing operations. If they held a degree in science or engineering, five years of practical experience were required.[37] While not specified in the constitution, the Membership Committee of the Institute tended to interpret the practical experience required of college graduates as being experience in "responsible charge."[38]

The Criteria Were Exclusive

These severe membership criteria meant that the early AIChE was a rather exclusive organization. The criteria excluded, for example, most chemists. In 1912 Institute President L. H. Baekeland noted that he had recommended several of his "chemical friends" for membership, only to find out that they were excluded by the constitution's requirement of proficiency in some branch of engineering as applied to chemical problems.[39] The requirement of engagement in chemical engineering practice at the time of election precluded most college faculty from becoming full members. Only a handful of academic chemical engineers—those who had long service in industry prior to teaching or were heavily involved in consulting—were eligible to join.[40] Thus the organization was at first largely composed of chemical manufacturers, chemical engineers in upper management, and consultants.

The organization's early exclusivity served several functions. First, it made the new society attractive to chemical executives and consultants discontent with the ACS' academic dominance and lack of special qualifications for membership.[41] At the same time, however, it minimized the new society's threat to the older organization's growth and programs. The requirements for joining the Institute were too stiff to draw many chemists from the general pool of ACS membership, and the AIChE requirement of proficiency and activity in some branch of engineering prevented analytical chemists working in industry, the core of the new Division of Industrial Chemistry and Chemical Engineering, from joining the Institute. And, as noted above, academic chemical engineers, one of the other sources of the older society's strength, were also largely excluded from the new organization.[42]

Other early AIChE policies made membership even more exclusive and further minimized competition between the two chemical societies. For many years it organized no membership drives, depending on personal contact alone to bring inquiries for membership.[43] Further, the Institute's Membership Committee interpreted the constitution's membership requirements with a bias toward quality instead of quantity. Samuel Sadtler noted in 1909: "This organization is something like a club, in that the Committee on Membership carefully scans qualifications."[44] Martin Ittner (President, 1936-37), a member of that committee for 20 years, noted that it had tended to view the constitutional requirements for membership as minimum requirements, not as a command to accept everyone who met the letter of the wording.[45] As a result the rejection rate was often quite high. Sometimes as many as 25% or more of those applying for full membership were turned down.[46]

Even the creation in 1910 of a second grade of membership (Junior) with lower membership requirements posed no threat to the vitality of the American Chemical Society. The new grade required a bachelor's degree or five years' experience in chemical technology plus active engagement in chemical engineering work.[47] But it enjoyed little popularity for many years. The grade attracted so few applicants that it was temporarily abolished between 1930 and 1932, and even as late as 1936 only 15% of AIChE members belonged to it.[48] It attracted so few for several reasons. First, the Institute made no serious effort to recruit younger engineers as members. Second, the grade offered few privileges. Junior members could not vote or run for office. They thus had little role to play in the Institute. Third, the Membership Committee applied very strict rules to applications for this grade as well as for the Active grade. According to Ittner, the Membership Committee not only looked at whether applicants for the Junior grade met the letter of the constitutional requirements, they also "weighed the probability of the applicant sticking to chemical engineering work and advancing sufficiently in the profession to qualify for active or associate membership."[49] The rate of rejection for applicants to Junior membership was, therefore, high.

In 1920 the AIChE did consider creating another class of membership—"Associate"— with reduced chemical engineering practice criteria to permit the admission of chemical engineering professors and industrial chemists with some interest in chemical engineering. A number of members recognized, however, that such a step could lead to direct competition with the ACS. Vice President Henry Howard, for instance, reminded members of the bitter feelings which had accompanied the birth of AIChE. He noted that the "whole excuse" for the Institute's existence was that it covered a field not filled by other organizations. Howard concluded: "I think we are dealing with something that is pretty dangerous to the future of the organization."[50] Persuaded by these and other arguments, the Institute wisely chose to avoid conflict. The plan to create an "Associate" member class open to chemists working in industry was dropped.

Largely as a result of its restrictive membership standards and the way these standards were applied, the AIChE grew slowly. There were 40 charter members in 1908 and 118 members by 1910. But there were only 344 members in 1920, when the creation of the "Associate" grade was debated; 644 in 1924; and 872 in 1930.[51] This growth rate may be compared with that of the Institute of Radio Engineers (IRE). Like the American Institute of Chemical Engineers, the early IRE had high membership standards, but it made a greater effort to attract younger engineers and interpreted its membership standards less strictly. The IRE was founded in 1912 with 46 charter members. In 1927 its membership was 4210; the AIChE's 723.[52]

By becoming, in the words of one of its early members, almost an "excursion club for a select few" in its early years,[53] the AIChE did not grow at a rapid rate, but it was successful in making membership attractive to those admitted, minimizing its threat to the American Chemical Society, and allowing the wounds and bitter feelings created by its secession to heal.

These wounds were further soothed by the rhetoric of early Institute leaders. For example, F. W. Frerichs (President, 1911), at a joint meeting with the ACS in December, 1910, argued that the two organizations should be considered as "supplementing one another in our aims," instead of as rivals. The AIChE's field, he declared, was "narrow" compared with the ACS', although "wide enough to support a profession which is represented by our Institute."[54] In 1911 Leo Baekeland (President, 1912) took a slightly different tack, emphasizing the continued "devotion" of AIChE members to the ACS.[55]

When direct confrontation between the two groups threatened, the AIChE in its early years acted with discretion. In 1914, for example, the Institute's committee on Chemical Engineering Education sought representation on a joint committee on engineering education organized by the other engineering societies, but found that ACS representatives had been admitted to that committee some years previously to provide input on chemical engineering education. The AIChE's intermediary recognized that the situation was "delicate" and pushed the matter slowly and cautiously, eventually securing representation on the committee without a messy confrontation.[56]

The success of the AIChE's decision to minimize conflicts by emphasizing its role as a select, complementary organization and by diplomacy is clear. Prior to the 1950s most Institute members were also members of the ACS. Browne and Weeks noted in 1951 that ACS members "still make up a great majority of its [AIChE's] membership."[57] Moreover, many chemical engineers served both societies well. A. D. Little, for instance, was President of the ACS in 1912-13 and President of the AIChE in 1919. Leo Baekeland was President of the ACS in 1924 and of the AIChE in 1912.

By treading carefully in its early years the AIChE gradually and peacefully secured recognition of its claims to represent the domain of chemical engineering. One example of this recognition was the decision of the American Chemical Society in September, 1940, to accept as its list of schools accredited for chemical engineering instruction the list of institutions accredited by the AIChE.[58] J. V. N. Dorr summed up the Institute's accomplishment in his 1932 Presidential Address:

> When our Institute was first formed there was a question in the minds of many members of the Chemical Societies then existing whether we were not further dividing their functions. There seems to be no question that during the past 25 years we have found a useful place and have been accorded a position that no other society would attempt to fill.[59]

CHEMICAL ENGINEERING EDUCATION

To become more than an exclusive, splinter society, the American Institute of Chemical Engineers had to do more than merely co-exist with the American Chemical Society. The legitimacy of chemical engineering as an independent discipline, rather than merely a hybrid of existing disciplines or a branch of industrial chemistry, had to be established. This problem was recognized early. F. W. Atkinson, President of Brooklyn Polytechnic Institute, noted in 1910: "Chemical engineering needs to be more sharply defined. Its scope is still in a somewhat indeterminate state and as yet its position as one of the professions is not clearly recognized."[60] H. O. Chute, a chemical engineer with over a dozen patents in the area of wood and grain distillation, put it differently in 1921: "We are not even able to convince other engineers that we are engineers."[61]

The effort to establish the scope and nature of chemical engineering and demonstrate its legitimacy as an independent profession focused initially on working out methods for reproducing practitioners, that is, on education. As Guédon has pointed out, the creation of a good, viable, standardized curriculum for producing the next generation of chemical engineers had the potential of making two important contributions to establishing chemical engineering as a new and viable discipline: (1) it would help define the disciplinary structure still lacking in 1908 and (2) it would help convince both industrialists and educators that chemical engineers were needed.[62]

The importance of propagating chemical engineering through education was recognized by the founders of the AIChE. In the June, 1908, address in which he sought to justify the creation of the Institute Charles McKenna declared:

> But the noblest aim before us, gentlemen, the one which most amply justifies us before all the world, is our ambition for the enlightenment and ample equipment of our successors: that is for the improvement of the training of the chemical engineer of the future.[63]

One of the first standing committees established by the new organization was a Committee on Chemical Engineering Education.

One of the major elements which distinguishes the AIChE's early history from that of the other major American engineering societies is the high emphasis which the Institute placed on educational activities due to the necessity of establishing the legitimacy of chemical engineering as an independent discipline. Practically all of the other early major American engineering societies—the American Society of Civil Engineers (ASCE), the American Institute of Mining and Metallurgical Engineers (AIME), the American Institute of Electrical Engineers (AIEE), the American Society of Mechanical Engineers (ASME), the Institute of Radio Engineers (IRE)—focused in their early years almost entirely on social and technical functions, on providing a means for their members to meet and exchange technical information. Though occasionally interested in educational matters, they did not play any major role in shaping the course of education in their fields. To some extent this was because, for them, it was not necessary. An academic structure had emerged for teaching civil, mechanical, and mining engineering before the emergence of their respective professional societies. While this was not as much the case with electrical and electronic engineering, they were distinct enough from most existing branches of engineering and science that establishment of a formal structure for training the next generation of practitioners presented only a few major problems.

Chemical engineering was different. Its domain was, as we have seen, strongly challenged on one side by chemistry and on the other side by mechanical engineering. To establish its position as an independent discipline, distinct from both chemistry and mechanical engineering and not merely a hybrid of those fields, the founders of AIChE had to create a curriculum distinct from the industrial chemistry, applied chemistry, and mechanical engineering traditions which claimed its subject area. Thus, far earlier than most engineering professional societies, the Institute became deeply involved in education.

Its interest in chemical engineering education came at a time when engineering education generally and chemical engineering education in particular were in a state of "flux."[64] The traditional method of training engineers—apprenticeship—which had placed the burden of education on the employer, was collapsing before the rise of science-based engineering. At the same time, the traditional university concentration on the liberal arts and pure sciences was crumbling before demands for a "practical" education and the growing scale and importance of American industry. The problem faced by all disciplines, but more critical for chemical engineering than most others, was how to work out a suitable division of labor between university and industry.[65]

Programs labeled "chemical engineering" had sprung up as early as 1888 at the Massachusetts Institute of Technology. By 1908 there were over a dozen. These curricula had little in common. The University of Pennsylvania's course, organized in 1892, consisted mostly of mechanical engineering (17 courses) with some chemistry (6 courses), but no courses labeled "chemical engineering." Other schools like MIT and Tulane established curricula primarily oriented toward industrial or applied chemistry, instead of mechanical engineering, but likewise offered no courses specifically labeled as "chemical engineering." At Wisconsin the Chemical Engineering Department emerged from the Electrical Engineering Department and had an electrical engineering bias. Other schools like Pratt Institute experimented with a near apprenticeship organization. Things were,

THE STRUGGLE FOR LEGITIMACY

indeed, in a state of flux, and opinions on how the technical needs of the chemical industry should be met through education varied from one extreme to the other.[66]

Believing that existing educational practices did not prepare students for work in the chemical industry and that universities and colleges did not know what industry required of them, the AIChE established a Committee on Chemical Engineering Education at its first regular meeting in Pittsburgh in December, 1908. It was charged with determining, through discussion and questionnaire, just what the education of a chemical engineer should include. This committee was the most dynamic and innovative element within the Institute over most of the next two decades and made it a major forum for discussing and exploring various experiments in chemical engineering education and various views on how chemical engineers should be trained.[67]

Consensus on such issues, however, proved initially difficult to obtain. The committee reported in December, 1913, that the opinions expressed by those responding to its early reports and canvasses were "so varied and often apparently mutually antagonistic" that it was unable to formulate specific resolutions."[68]

Nonetheless, the Education Committee was able to reach consensus on some general principles. It suggested that curricula which focused on or attempted to provide practical experience in specific industries had no place in university education. Provision of practical experience had to be left to the employer. The school's duty was to provide graduates with a thorough grounding in the basic principles of chemistry and engineering.[69] This particular division of labor was somewhat more explicit than earlier suggestions, but it was not novel, for it had been suggested several times in, for example, a 1904 ACS discussion on the education of technical chemists.[70] The key problem, however, was transforming general principles into a specific curriculum distinct from existing subject fields like chemistry and mechanical engineering.

To determine what universities were doing to prepare chemical engineers, the Committee on Engineering Education recommended in 1914 that the Institute undertake a detailed study of the "chemical engineering" curricula at all institutions offering courses under that name. It suggested that a member of the committee spend a week or two at each institution gathering data which would be compiled into a detailed, comparative report. Such a report, the committee hoped, would provide support for the view that chemical engineering was developing as a distinct discipline and not merely a hybrid of chemistry and mechanical engineering.[71] But in 1914 the AIChE scarcely had 200 members and could ill afford such an ambitious study. The project was postponed.

Between 1914 and 1919 or 1920 the Institute was relatively inactive in education. In part this was due to the Carnegie Foundation's plans for a comprehensive study of engineering education in cooperation with the Society for the Promotion of Engineering Education. The foundation had no definite plans to include a new and seemingly minor field like chemical engineering in its study. Thus the focus of the Education Committee's activities was diverted from its own plans to attempting to persuade the foundation to include chemical engineering in its study. By 1916 it had succeeded in this task and went dormant while awaiting the outcome of this study.[72] World War I further encouraged this dormancy. The committee's chairman observed in 1919 that the war had so absorbed the "interests and energies" of members that little enthusiasm had been left for the Institute's educational work.[73]

In 1919, however, this work was revived. The Carnegie Foundation's report (known as the Mann Report), issued in 1918, had provided little assistance to chemical engineers attempting, through the university curriculum, to establish the scope and nature of the profession. It had concentrated on the nature of engineering education generally and had dealt with subjects like the length of courses, the amount of preparation required, and the necessity of cultural studies.[74]

In the meantime, the need for some standardization in chemical engineering courses was becoming steadily more critical. As a result of the suspension of German exports in the fine chemicals field during World War I, the American chemical industry's growth rate had accelerated. In the wake of this expansion the number of chemical engineering students had multiplied rapidly. In 1909 less than 1000 students were enrolled in chemical engineering. By 1920-21 enrollment had reached almost 6000.[75] The variety of curricula, their content and requirements, had, if anything, only grown more diverse since 1914 when the Chemical Engineering Education Committee had outlined the basic principles which it felt should guide the education of chemical engineers.

Unit Operations Arrived

In the same interval, however, the key to transforming these basic principles into concrete form had emerged—the concept of "unit operations." This concept was first explicitly stated by A. D. Little in 1915 in a report to the President of the Massachusetts Institute of Technology:

> Any chemical process, on whatever scale conducted, may be resolved into a coördinate series of what may be termed 'Unit Operations,' as pulverizing, dyeing, roasting, crystallizing, filtering, evaporation, electrolyzing, and so on. The number of these basic unit operations is not large and relatively few of them are involved in any particular process. The complexity of chemical engineering results from the variety of conditions as to temperature, pressure, etc., under which the unit operations must be carried out in different processes, and from the limitations as to material of construction and design of apparatus imposed by the physical and chemical character of the reacting substances.[76]

The "unit operations" approach sharply delimited the domain of chemical engineering and clearly distinguished chemical engineering from industrial chemistry, applied chemistry, and mechanical engineering. The industrial chemist had concentrated on product; the applied chemist on individual reactions; the mechanical engineer on machinery. None of these had focused on processes, and none had clearly recognized the "unit operations" common to a variety of chemical products, chemical reactions, and chemical machinery. Little sensed that each of the unit operations could be approached just as quantitatively and systematically as individual chemical reactions or pieces of machinery and studied as special sciences. In a sense what Little and his associates had done was to recognize and make explicit the alphabet out of which the language of chemical engineering was to be written.[77]

Because chemical engineering curricula had grown without direction and, perhaps, because some members felt that unit operations had clarified the scope and nature of chemical engineering, at the December, 1919, meeting of the AIChE, the Committee on Chemical Engineering Education was asked to begin considering steps which the organization could take toward increasing the competency of chemical engineering graduates.[78]

Under A. D. Little's chairmanship, the Committee on Chemical Engineering Education now resumed activity, preparing in 1920 an extensive bibliography on the education and training of chemists and chemical engineers and corresponding with 128 schools offering or believed to be offering courses relevant to chemical engineering. Little noted in 1920 that his committee had already compiled data on the courses being offered in chemical engineering in 53 schools.[79] By the end of the year it had identified 77 institutions offering chemical engineering courses and had begun to analyze and tabulate data from them. Preliminary findings indicated a "bewildering" number of subjects being required in chemical engineering curricula, "great variations" in the relative weight given courses, and "little evidence" of anything like a standard course. Little concluded that a radical change in the organization of courses and teaching methods "appears to be desirable, if not imperative."[80]

In 1921-22 the committee studied in greater detail 78 institutions offering programs labeled "chemical engineering," analyzing their curricula and developing recommendations. An interim report was made to the Institute in 1921, a final report in 1922.[81]

The final report pointed out in detail the "wide divergence" in nomenclature, "the absence of anything approaching uniformity" in weighing and in arranging courses, and the need for some standardization. Not surprisingly the committee rejected the premise that chemical engineering was a hybrid of various engineering fields and chemistry and accepted unit operations as a means of defining the profession's boundaries. It declared:

> Chemical engineering . . . , as distinguished from the aggregate number of subjects comprised in courses of that name, is not a composite of chemistry and mechanical and civil engineering, but a science of itself, the basis of which is those unit operations which in their proper sequence and coördination constitute a chemical process as conducted on the industrial scale.[82]

To bring chemical engineering education into line with this conception of the discipline the committee recommended reduction in the multiplicity of subjects included in existing curricula; omission of civil, sanitary, industrial, mining, and electrical engineering courses beyond the fundamentals; the avoidance of specialization by industry; standardization of nomenclature and weighing of chemical engineering courses; and the provision of a strong background in basic chemistry, physics, mathematics, and chemical engineering to students. Several additional recommendations concerned teaching practices and admission standards.

Because no authority existed to implement these recommendations and because the committee felt that the need for reorganization of chemical engineering courses was "evident and urgent," it suggested that the AIChE sponsor a meeting with representatives from educational institutions to discuss the report.[83]

A Monumental Report

The Institute accepted the report of Little's committee enthusiastically. President David Wesson called it "monumental." Ralph McKee (Director 1923-25) observed: "The Committee have written a prescription, and it is our duty to see that the prescription is filled at the drug store, and given to the patients, and to see also if we can get the patient to use the data which we have gathered together." The Institute voted to accept the report and to have it printed and distributed to interested parties.[84]

At Little's recommendation, the AIChE also called a conference with educators at Brooklyn Polytechnic on May 16, 1922, to discuss the report. At this meeting some minor changes were made in the recommendations, but the original report was largely approved and the meeting was considered "very successful."[85] At its Niagara Falls meeting in June, 1922, the AIChE accepted the revised report and appointed a new Committee on Chemical Engineering Education composed of five members from industry and five from education, plus the chairman, H. C. Parmelee (Secretary 1927-29), formerly Chief Chemist of the Globe Plant of American Smelting and Refining Co. and a consulting chemist and, at that time, Chief Editor of *Chemical and Metallurgical Engineering*. This committee was instructed to continue studying various chemical engineering curricula, to attempt to persuade institutions teaching chemical engineering to accept the conclusions of the Little report, and to publish, after a three-year period, the names of those institutions whose programs in chemical engineering could be considered as satisfactory.[86]

Between 1922 and 1925 the engineering education committee worked at these tasks. It developed a classification for courses offered in chemical engineering curricula, set up a tentative norm for comparison, and reviewed the offerings of a large number of schools. It also began to establish preliminary standards for chemical engineering departments in such areas as entrance requirements, time required for completion of courses, organizational autonomy, faculty qualifications, and physical facilities. The committee summarized this work in its June, 1925, report and included a list of 14 schools regarded as meeting its standards. In addition, it recommended that the AIChE establish a permanent standing committee to periodically review schools and add to this list.[87]

The report was accepted by the Institute, but with some dissension over one of its recommendations—that departments of chemical engineering be administered within colleges of engineering instead of departments of chemistry. Chemical engineers with backgrounds in chemistry feared that this would lead to a dilution of chemistry requirements and moved to amend the report to eliminate the provision. In response Parmelee argued that if chemical engineering was simply applied chemistry then the Institute should "wind up" its affairs and join the American Chemical Society. John Olsen, the Institute's Secretary, backed Parmelee, noting that chemical engineering had only made progress in his institution after it had been removed from the chemistry department. The move to eliminate the recommendation that chemical engineering departments be administered in colleges of engineering was defeated and the report adopted in its original form.[88]

The Institute's decision to accredit curricula was not, however, without opposition. A 1927 questionnaire indicated that 30.5% of the membership did not wish AIChE to be involved in accreditation activities.[89] Furthermore, some who approved accreditation generally opposed specific elements of the program, for instance, as noted above, the emphasis on administration within engineering rather than chemistry.[90] In March, 1927, Council reviewed the organization's policy of endorsing curricula but concluded that "in spite of unfavorable comment the Institute has embarked on a policy which ought to be continued."[91] By 1931, 18 schools had been accredited,[92] and accreditation had become a regular function of the Institute.

Because AIChE accreditation standards emphasized the importance of unit operations as the core of chemical engineering curricula and favored the creation of independent departments in engineering schools rather than sub-departments within chemistry departments, the Institute put pressure on chemical engineering programs all over the country to modify their organizations and offerings in a manner that clearly delineated the boundaries of chemical engineering and promoted its position as an independent discipline. In 1928 Alfred Holmes White, a prominent educator and future (1929-30)

THE STRUGGLE FOR LEGITIMACY

AIChE President, noted: "almost all schools which teach chemical engineering now recognize these unit processes [i.e., "unit operations," see note 93] as providing the framework for the engineering side of chemical engineering."[93]

The AIChE's decision to monitor and direct educational institutions through campus inspection visits and accreditation was completely unprecedented in the engineering community. Other engineering societies had occasionally talked about regulating education (for instance, ASCE in 1874)[94] but none had done anything, either because their need to delineate the boundaries of their profession and to develop a means of reproducing practitioners was not as critical or because they believed that educational institutions would not allow professional societies to dictate the structure of their programs.[95]

Only in the early 1930s, as a result of AIChE's success in the area, did other major engineering professional societies organize to accredit programs in their disciplines, forming in 1932 the Engineers' Council for Professional Development (ECPD), now called the Accreditation Board for Engineering and Technology (ABET). Although not considered at the time to be in the same league as the "Founder Societies" (American Society of Civil Engineers; American Society of Mechanical Engineers; American Institute of Electrical Engineers; American Institute of Mining and Metallurgical Engineers) which, together with the Society for the Promotion of Engineering Education and the National Council of Engineering Examiners, established ECPD, AIChE was invited to join the organization since it had been the pioneer in the area of accreditation.

AIChE agreed to join in designating ECPD as the accrediting agency for engineering curricula, but it insisted on, and was granted, a somewhat special status because it was the only society with an accreditation scheme already in operation and because it demanded higher standards for accreditation than the other societies. For instance, AIChE alone expected engineering faculty to be engaged in research. Under the agreement by which AIChE joined ECPD, AIChE's Education Committee was designated as the subcommittee in charge of chemical engineering programs of the ECPD's Committee on Engineering Schools. When schools requested accreditation, the chemical engineering member of the visiting committee was granted the special privilege of not reporting directly back to the ECPD Committee on Engineering Schools, but, instead, to AIChE's Committee on Chemical Engineering Education, which would independently submit its recommendation to ECPD.[96] This agreement is still in force.

At the AIChE's 25th anniversary in 1933 the Committee on Chemical Engineering Education could look back on its accomplishments with considerable satisfaction. H. C. Parmelee, who summed up the committee's work in the silver anniversary volume of the American Institute of Chemical Engineers' *Transactions,* was correct in declaring that its work was "outstanding among the Institute's activities in the past twenty-five years."[97] The society's educational work had achieved more tangible results than its work in any other area. As a result of the Institute's efforts, chemical engineering by the late 1920s had achieved a position in educational institutions as an independent and well-recognized division of engineering,[98] the chaotic curricula offered in the early years had been standardized, and the practice of accreditation had been accepted by the entire engineering profession.

OTHER EARLY INSTITUTE PROGRAMS

Although the reorganization of educational curricula around the concept of unit operations was the single most important factor in the establishment of chemical engineering as a legitimate and distinct profession, the efforts of the Institute to promote and diffuse technical information through its publications and meetings were also important, for the primary criterion of a profession is its possession of a unique body of "esoteric knowledge."[99]

Publications and Meetings

At the founding many academicians feared that, despite good intentions, a professional chemical engineering organization would not be able to perform the traditional functions of a professional scientific or engineering society—the promotion and dissemination of knowledge. At the January, 1908, meeting of the Committee of Fifty several who opposed the creation of the AIChE argued that due to industrial secrecy, it would be almost impossible to obtain good chemical engineering papers.[100] In 1932 John Olsen, the man most familiar with the Institute's operation over its first quarter century, having served as Secretary from 1908 to 1927 and as President in 1931, confessed that this objection had "some validity."[101] There was, moreover, an additional problem. As one of the early members of the Institute noted, in 1908 there was "hardly a trace" of an organized body of knowledge which could be called chemical engineering and no clear understanding of what the field should cover.[102]

Undaunted, the Committee of Six recommended in early 1908 the regular publication of transactions, and the publication and distribution of papers which increased knowledge of chemical engineering was made one of the constitutional objectives of the new organization.[103] Moreover, Samuel Sadtler in his Presidential Address of December, 1908, described the publication and distribution of chemical engineering papers as one of the most important of AIChE's objectives.[104]

The decision to publish only transactions and not to undertake the publication of a broader journal which would include industrial news and papers not presented at AIChE meetings was conservative. But there was probably good reason for caution. A more ambitious publications program would have seriously taxed the very limited resources of the new organization and might well have thrown it into direct competition with the ACS which launched *I&EC* at about the same time the Institute was organized.

The early papers published in the Institute's *Transactions* reveal the same problems that early chemical engineering curricula suffered: a lack of focus, a lack of firm understanding of where the boundaries of the new discipline were and what separated it from industrial chemistry, and ignorance of what areas needed research. Frederick Zeisberg (President, 1938), Chairman of the AIChE Publications Committee in 1924 and from 1927 to 1932, observed that the early volumes contained papers on industrial chemistry and other subjects which could no longer be considered as belonging in the domain of chemical engineering.[105]

The emergence and acceptance of unit operations as a means of delimiting the boundaries of chemical engineering had an important influence on the nature and quality of papers published by the Institute, as well as on chemical engineering education. This impact is reflected in Zeisberg's observation that about 1919 the nature of papers presented to the AIChE began to change. He noted that papers simply describing chemical production began to decline in number, while those providing quantitative data that could be applied to chemical plant design began to appear with increasing frequency.[106]

Although for nearly 40 years the *Transactions* were the Institute's primary contribution to chemical engineering literature, it did attempt several other ventures in its early years. In 1910, recognizing the need for a readily available, centralized source of information on chemical engineering equipment, it appointed a Chemical Engineering Catalog Committee. Over the next few years this committee established general guidelines for the proposed catalog and attempted to find a publisher. An initial attempt to get the F. W. Dodge Co. to publish the compilation failed. But in 1915 the Chemical Catalog Co. took up the project at AIChE's urging. The Institute's committee provided the publisher with considerable supervision in the early years of the catalog's production, reducing its efforts as the catalog became established. In 1955 the committee, having long since discharged its function, was disbanded.

The Institute also played a major role in another publication venture, this one having utility beyond the limits of chemical engineering. The *International Critical Tables* project of the National Research Council was neither originated by nor initially sponsored by AIChE. But by 1920 the Institute had taken the major role in providing impetus for the project. As then AIChE President H. K. Moore observed:

> While this was instituted by another organization, it has been largely the effort of your officers, and of this Institute, that has put this across. It would have been a failure if it had not been for the work which our members and officers put into this.[107]

In the mid to late 1920s the Institute's publication program encountered difficulties. The long lag between when papers were read and when the annual transactions appeared (a lag at least partially due to reliance on a part-time secretary for most editorial and publication work) and their limited circulation made other publications more popular and made it increasingly more difficult for the Institute's Committee on Publications to secure good papers.[108] Thus many of the important contributions to the literature of chemical engineering, particularly in the 1920s, were published by AIChE members, but outside of the *Transactions*. They appeared in other journals, as well as in monographs and textbooks.

The most important of the latter was *Principles of Chemical Engineering*, published in 1923 and authored by William H. Walker, W. K. Lewis, and W. H. McAdams of MIT. This popular text was the first to utilize unit operations as the scientific basis for chemical engineering. According to one of the early historians of the AIChE, it "afforded tangible proof that chemical engineering had achieved professional status in its own distinct field."[109] Walker, the principal author of the Principles, had been one of the Committee of Six which had presided over the foundation of the Institute.

Because personal contact is often more important (though harder to document) than published papers in the exchange of technical data and ideas, the AIChE's meetings program was probably just as important as its early publications program in promoting the growth of the new discipline. For some 35 years after its founding the society held two meetings a year: one in the late spring or early summer, another in the latter part of the year. The papers presented at these meetings were solicited, until 1932, by the Publications Committee, largely on the basis of suitability for ultimate publication in the *Transactions*. Often, however, the plant tours scheduled in conjunction with the meetings attracted more enthusiasm than the papers.[110]

Meetings, publications, and accreditation activities drew heavily on the slender resources of the early AIChE and limited the programs it could undertake in other fields. Nonetheless, the organization's performance in several of these fields, for example ethics and governmental interaction, was at least respectable.

Code of Ethics

In the early twentieth century many American professional organizations adopted formal codes of ethics, some as a means of actually trying to regulate the manner in which the profession was practiced, others as a means of demonstrating social concern or achieving the high professional status accorded law, medicine, and the clergy. The American Society of Civil Engineers discussed adopting a code as early as 1893, and again in 1902, but adopted one on neither occasion.[111] In 1906 the president of the American Institute of Electrical Engineers, Schuyler Wheeler, called on that organization to adopt a code of ethics. In 1907 a code was written and submitted to the membership for approval, but not adopted. The AIEE formally adopted a code only in March, 1912. In the meantime the newly-organized American Institute of Consulting Engineers had adopted, in May, 1911, the first code of ethics for an American engineering organization.[112]

The American Institute of Chemical Engineers began considering a code of ethics seven months later. At the December, 1911, meeting of the society, A. C. Langmuir (Vice President 1914–18) brought up the subject, pointing out that the American Chemical Society was considering such a code. The matter was referred to a committee of three which included Langmuir.[113] This committee presented a draft of a code to the September, 1912, New York meeting of the Institute, where it was amended.[114] Formal adoption came at the December, 1912, Detroit meeting.[115] Thus, AIChE had a code only four years after its founding, a most notable accomplishment.

The Institute's code was worded in very general terms, enunciating only general guidelines rather than defining specific actions as ethical or unethical. It included no enforcement machinery. And its provisions tended to apply much more realistically to independently-employed professionals, like consultants, than to engineers employed in industrial bureaucracies. The codes adopted by other engineering societies in the era had much the same characteristics.[116] The AIChE's code was seldom used to censure members. It was used early in the Institute's history, however, to screen out applicants for membership. John Olsen reported in 1920 that about one in every 10 or 11 applicants was denied membership on ethical grounds.[117]

Government Interaction

At its birth some members wanted the AIChE to be politically active. For example, in justifying the creation of the organization in June, 1908, McKenna argued that chemical engineering should have more input into public policy matters.[118] One of the constitutional objectives of the Institute was giving the profession sufficient standing so that its advice would be sought and heeded by governmental bodies.[119]

These desires were not novel. In the 1880s the American Society of Mechanical Engineers had launched a political action program. But it had found that it was extremely hard for an engineering society to exert any significant influence on Congress, that engineers did not function well in the political arena, and that most of its political activities either aroused little enthusiasm among its membership or left them disgusted.[120]

The AIChE's political action program was less well directed and less ambitious than the early ASME's. But the Institute did attempt to influence legislation in a number of areas. In 1911 concerns about the patent system led the AIChE to appoint a Committee on Patents. In 1912 the Institute sent three delegates to provide for a chemical engineering input at a Washington conference on patent problems and began a publicity campaign for patent reform.[121] Between 1918 and 1922 the AIChE had a metric committee which monitored legislation pertaining to the metric system and pushed for adoption of the system.[122] In 1919 the Institute passed several resolutions, forwarding them to the President and Congress. One urged tariff protection of the chemical industry; another urged governmental bodies to pay their bills on time; a third condemned the Anti-Dumping Law as ineffective.[123] In 1921 the Institute passed resolutions favoring the creation of a Department of Public Works.[124] In 1924 it created a Committee on Industrial Alcohol to speak for the Institute on matters concerning the use and sale of alcohol.[125] In addition, the AIChE passed at various times resolutions urging Congressional action on a variety of other subjects: establishment of a single court of patent appeals, addition of a working clause to the American patent system, institution of a tariff system to protect the American coal tar dye industry, and maintenance of the Chemical Warfare Service, for example.[126]

But AIChE, like ASME, found governmental involvement fraught with problems. A good example is the issue of patent reform. In June, 1911, the Institute appointed its Committee on Patents. This committee investigated what other societies had done in the area and attempted to draft a resolution calling on Congress to investigate and reform the patent system. It quickly discovered that political affairs were highly divisive. There was "severe" opposition even to a simple resolution calling for study and revision of the patent system from around 10% of those it questioned. Even within the three-man committee there was little consensus about what specifically was wrong with the patent system or the best way to correct the admitted problems. The committee also discovered that other larger, better financed, and more powerful groups had been striving for years to influence Congress on patents with little effect. Faced with these discoveries it simply recommended that the Institute sponsor a symposium on the American patent system.[127]

Despite these early frustrations, the Institute's patent committee continued for several years trying to persuade Congress to reform the patent system, pushing, in particular, the idea of a single court of patent appeals, and even attempting to launch in 1912–13 a campaign to influence public opinion on the issue.[128] But the results of these efforts were disappointing. W. M. Grosvenor, a former chairman of the committee, reported in 1914 that the Institute's recommendation of a single court of patent appeals "seems to find general endorsement, but is always sidetracked." Grosvenor added: "There is practically nothing we can do to definitely bring this to a head."[129] Maximilian Toch, long active in patent activities and Chairman of the Committee on Patents from 1927 to 1929, reported in 1919 that it was "absolutely impossible to get any legislators to listen to anything concerning patents or the metric system."[130] Leo Baekeland, a former AIChE President and Chairman of the Patent Committee from 1913 to 1926, was similarly frustrated, noting that a patent reform bill he supported had been killed at the last moment in a recent session of Congress, making "all our work ... useless." "All this," he added, "is a rather discouraging result of many years' efforts."[131]

The Institute's Committee on Patents thus found, as the ASME had discovered some 40 to 50 years earlier, that attempts by engineering societies to influence the legislative process were usually frustrating and disappointing. The Patent Committee survived until 1953, but beginning in the 1920s is served mainly as a watchdog, informing the Institute of pending patent legislation, occasionally urging members, as individuals, to write their congressmen about some particularly objectionable piece of legislation. The Patent

Committee was disbanded in 1953 when its chairman argued that the committee's functions were being more effectively served by the patent section of the American Bar Association.[132]

The AIChE's Committee on Industrial Alcohol, appointed in 1924, encountered similar difficulties. It was effective in monitoring and informing members of pending bills affecting the supply of industrial alcohol, but found itself unable to influence Congress in a significant manner. The committee's chairman reported in 1926 that all efforts to have alcohol legislation "considered on its actual merits" had been "futile."[133]

Thus AIChE's activities in the governmental sphere in the period from 1908 to 1930 were neither unusual nor effective. Even within the Institute's limited and exclusive membership it was sometimes difficult to get agreement on specific actions, and even when agreement was possible the organization's impact on governmental policy was clearly quite limited. Not surprisingly, then, the Institute's activities in the political arena attracted neither the same support nor enthusiasm as its much more effective performances in education, publications, and meetings.

SUMMARY

Despite the futility of its forays into the governmental arena, the AIChE achieved a respectable record in the first 20 to 25 years of its existence. It had established modestly successful meetings and publications programs and a very successful accreditation program. It had undertaken, in addition, a variety of non-technical activities including adoption of a code of ethics and attempts to influence legislation. It had, also, maintained a slow but steady rate of growth. By 1930 AIChE membership had reached 872. As a result of this growth in programs and in numbers AIChE's Council considered opening a permanent headquarters with a paid Executive Secretary several times in the late 1920s and approved the idea in 1929.[134] These headquarters were opened in Philadelphia in February, 1930.

Founded at a time when the very term "chemical engineering" was vague and often misunderstood, the AIChE had played a major role in defining the profession and delimiting its boundaries. Sidney Kirkpatrick (President, 1942) observed in the early 1930s that the Institute had achieved recognition for being "as important in its field as are the four older and so-called Founder Societies of mechanical, civil, electrical and mining engineers."[135] And J. V. N. Dorr, 1932-33 President, commented: "The four founder societies while tending originally to regard us as 'queer kinds of chemists' now recognize us as a fifth classification of basic engineering."[136] Chemical engineering had, in other words, been established as a legitimate discipline by 1930.

Chapter 1: Notes

1. For the histories of these societies see: William H. Wisely, *The American Civil Engineer, 1852-1974: The History, Traditions and Development of the American Society of Civil Engineers*, New York, 1974; *Centennial Volume: American Institute of Mining, Metallurgical and Petroleum Engineers, 1871-1970*, New York, 1971; and Bruce Sinclair, *A Centennial History of the American Society of Mechanical Engineers, 1880-1980*, Toronto, 1980.
2. F. J. Van Antwerpen, "The Origins of Chemical Engineering," in William F. Furter, ed., *History of Chemical Engineering*, Washington, 1980, pp. 1-14, and D. M. Newitt, "The Origins of Chemical Engineering," in Herbert W. Cremer, ed., *Chemical Engineering Practice*,

v. 1, London, 1956, pp. 1–22, review some of the early roots of chemical engineering. But even though some activities existed as early as the sixteenth century which could be called chemical engineering, they were not recognized as examples of chemical engineering until the emergence of the profession in the twentieth century.
3. The traditions which had claims to the realm of knowledge later controlled by chemical enginering are reviewed in more detail in an excellent article by Jean-Claude Guédon, "Il progett dell'ingegneria chimica: L'affermazione delle operazioni di base negli Stati Uniti," *Testi & Contesti*, v. 5 (1981) pp. 5–27, esp. pp. 8–12. I have utilized several ideas from this essay and hope that it will be published in English someday.
4. See Guédon, "Conceptual and Institutional Obstacles to the Emergence of Unit Operations in Europe," in Furter, ed., *History of Chemical Engineering*, pp. 45–75, esp. pp. 62–70. See also Klaus Buchholz, "*Verfahrenstechnik* (Chemical Engineering)—Its Development, Present State and Structure," *Social Studies of Science*, v. 9 (1979) pp. 33–62.
5. J. B. F. Herreshoff, "The Training of Technical Chemists," *Science*, v. 19 (1904) pp. 574 (Schweitzer), 575 (Bogert).
6. Manufacturing Chemists' Assn., *Chemical Facts & Figures: Annual Statistics of the American Chemical Industry*, n. p. [Washington], 1940, p. 1; L. F. Haber, *The Chemical Industry, 1900–1930*, Oxford, 1971, pp. 173–74.
7. Haber, *Chemical Industry*, pp. 173–74, and B. G. Reuben and M. L. Burstall, *The Chemical Economy*, London, 1973, p. 19.
8. Martha M. Trescott, "Unit Operations in the Chemical Industry: An American Innovation in Modern Chemical Engineering," in William F. Furter, ed., *A Century of Chemical Engineering*, New York and London, 1982, pp. 10, 16–17; Haber, *Chemical Industry*, p. 174.
9. *AIChE Transactions* (hereafter *Transactions*), v. 3 (1910) p. 129 (Letter from Charles McKenna to F. W. Frerichs); Trescott, "Unit Operations," pp. 10, 16–17; Haber, *Chemical Industry*, p. 176.
10. Guédon, "Conceptual Obstacles to Unit Operations," pp. 45–76.
11. Herreshoff, "Training of Technical Chemists," p. 568 (discussion).
12. *Ibid.*, p. 563.
13. *Ibid.*, p. 561.
14. For instance, Alfred H. White, "Chemical Engineering Education: Building for the Future of the Profession," in Sidney D. Kirkpatrick, ed., *Twenty-Five Years of Chemical Engineering Progress*, New York, 1933, pp. 355–56 (Silver Anniversary Volume, American Institute of Chemical Engineers).
15. Richard K. Meade, "Why Not 'The American Society of Chemical Engineers?'," *The Chemical Engineer*, v. 2 (1905) pp. 391–93. Reprinted in *Chemical Engineering Progress* (Hereafter *CEP*), v. 54 (May, 1958) pp. 52–53.
16. *The Chemical Engineer*, v. 5 (1907) pp. 227–28.
17. *Transactions*, v. 1 (1908) p. 1, for Meade's letter; also, John C. Olsen, "Origins and Early Growth of the American Institute of Chemical Engineers," *Transactions*, v. 28 (1932) p. 300.
18. Meade, "The American Institute of Chemical Engineers," *The Chemical Engineer*, v. 8 (1908) p. 1; "A Proposed Society of Chemical Engineers," *ibid.*, v. 6 (1907) pp. 43–45; Olsen, "Origin and Early Growth," pp. 302–03; *Transactions*, v. 1 (1908) pp. 1–3; and F. J. Van Antwerpen and Sylvia Fourdrinier, *High Lights: The First Fifty Years of the American Institute of Chemical Engineers*, New York, 1958, pp. 8–9 (Hereafter *High Lights*).
19. Meade, "American Institute of Chemical Engineers," pp. 1–2; Olsen, "Origin and Early Growth," pp. 303–06; *Transactions*, v. 1 (1908) pp. 1–3; *High Lights*, pp. 9–11.
20. Meade, "American Institute of Chemical Engineers," p. 2; *High Lights*, p. 10.
21. For reports of this meeting see: *Transactions*, v. 1 (1908) p. 4; Olsen, "Origin and Early Growth," pp. 306–10; *High Lights*, pp. 11–21, citing "Proceedings of a Meeting Held at Parlor A, Belmont Hotel, Saturday Evening, January 18, 1908, for the Purpose of Discussing the Advisability of Forming a Society of Chemical Engineers," privately printed, 1908; and Charles

Albert Browne and Mary Elvira Weeks, *A History of the American Chemical Society,* Washington, 1952, pp. 85-86.
22. Meade, "American Institute of Chemical Engineers," p. 2; *High Lights,* p. 21.
23. *Transactions,* v. 1 (1908) p. 5; Olsen, "Origin and Early Growth," pp. 310-11.
24. "American Institute of Chemical Engineers," pp. 2-3; *High Lights,* p. 23.
25. Various sources give different figures on attendance at the Philadelphia meeting. The official minutes, however, list only 19 in attendance. The minutes are reprinted in *CEP,* v. 54 (May, 1958) p. 57, and *Transactions,* v. 1 (1908) pp. 6-7.
26. Charles F. McKenna, "The Justification of the American Institute of Chemical Engineers," *Transactions,* v. 1 (1908) pp. 8-20; portions reprinted in *CEP,* v. 54 (May, 1958) pp. 58-60.
27. For the Philadelphia meeting see: *High Lights,* pp. 23-25; Olsen, "Origin and Early Growth," pp. 311-14; *Transactions,* v. 1 (1908) pp. 6-7; "American Institute of Chemical Engineers," pp. 3-20; and *CEP,* v. 54 (May, 1958) pp. 56-57.
28. The AIChE's first constitution is printed in *Transactions,* v. 1 (1908) pp. 21-26. Edwin T. Layton, Jr., *The Revolt of the Engineers: Social Responsibility and the American Engineering Profession,* Cleveland and London, 1971, p. 15, notes that historically American engineering societies have not been openly democratic. The AIChE's early electoral system most closely paralleled that of the American Institute of Electrical Engineers, which, according to Layton (p. 81) had a "unique system of open and competitive elections."
29. Browne and Weeks, *History of the American Chemical Society,* pp. 68-96 passim.
30. *Ibid.,* pp. 70, 76.
31. "American Institute of Chemical Engineers," p. 2; *High Lights,* pp. 12-13, 15, 21; Browne and Weeks, *History of the American Chemical Society,* p. 89.
32. Browne and Weeks, *History of the American Chemical Society,* pp. 84-85, 89.
33. *AIChE Bulletin,* (Hereafter, *Bulletin*), no. 22 (1920) p. 49.
34. Browne and Weeks, *History of the American Chemical Society,* p. 86.
35. *Ibid.,* p. 85, 266, 370.
36. *Ibid.,* p. 88.
37. *Transactions,* v. 1 (1908) pp. 21-22, for membership requirements in first constitution.
38. *Ibid.,* v. 2 (1909) p. 23.
39. *Bulletin,* no. 6 (1912) p. 17.
40. *Ibid.,* no. 15 (1917) p. 6; no. 17 (1918) p. 21; no. 20 (1919) pp. 7-8, 13-14, 17-18.
41. Browne and Weeks, *History of the American Chemical Society,* pp. 86, 88; Guédon, "Progetto dell'ingegneria chimica," pp. 14-16, emphasizes the importance of the AIChE's restrictive membership criteria in reducing conflict with the ACS.
42. See, for example, *Transactions,* v. 4 (1911) pp. 65-66, for discontent with academic dominance of ACS.
43. For instance, *Bulletin,* no. 28 (1923) p. 17, where the advantages of a membership drive were considered, but rejected.
44. *Transactions,* v. 2 (1909) p. 6.
45. Martin H. Ittner, "The Membership Committee," *Transactions,* v. 28 (1932) pp. 317-18.
46. *Bulletin,* no. 21 (1920) p. 14, for instance.
47. *Transactions,* v. 2 (1909) pp. 19-23; v. 3 (1910) p. 40.
48. *High Lights,* front and rear covers, provide a table of the annual growth of AIChE membership from 1908 to 1958. I have relied on it for all pre-1958 membership statistics.
49. Ittner, "Membership Committee," p. 317.
50. *Bulletin,* no. 22 (1920) pp. 47, 49.
51. See note 48 above.
52. Laurens E. Whittemore, "The Institute of Radio Engineers—Fifty Years of Service," Institute of Radio Engineers, *Proceedings,* v. 50 (1962) p. 534.
53. *Bulletin,* no. 22 (1920) p. 39.
54. *Transactions,* v. 3 (1910) pp. 65-67.

55. *Ibid.*, v. 4 (1911) p. 56. For similar assertions of devotion to ACS see *ibid.*, v. 4 (1911) p. 65, and *Bulletin*, no. 14 (1917) p. 41.
56. *Transactions*, v. 9 (1916) pp. 23-24; *Bulletin*, no. 11 (1915) pp. 16-17.
57. Browne and Weeks, *History of the American Chemical Society*, p. 88. In 1951 this was not literally true. According to *CEP*, v. 49 (November, 1953) p. 23n, only 51% of AIChE members belong to ACS, down from 59% in 1947. These figures do not represent a "great" majority. Today, according to J. Charles Forman, AIChE Executive Director, less than 10% of ACS members are chemical engineers.
58. Browne and Weeks, *History of the American Chemical Society*, p. 209.
59. *Bulletin*, no. 45 (1932) p. 49.
60. *Transactions*, v. 3 (1910) p. 156.
61. *Bulletin*, no. 24 (1921) p. 53.
62. Guédon, "Progetto dell'ingegneria chimica," pp. 15-16. Herbert Dow noted in 1921: "In order to make the subject of Chemical Engineering as great and valuable as possible, it must contain within itself the germs of growth and development. No method of education that will facilitate the acquiring of new ideas has yet been fully worked out." See *Bulletin*, no. 23 (1921) p. 19.
63. McKenna, "Justification," p. 18.
64. This term was used by John Olsen in *Transactions*, v. 2 (1909) p. 25.
65. Guédon, "Progetto dell'ingegneria chimica," p. 16.
66. The early history of chemical engineering education is reviewed by J. W. Westwater, "The Beginnings of Chemical Engineering Education in the USA," in Furter, ed., *History of Chemical Engineering*, pp. 141-52; Guédon, "Progetto dell'ingegneria chimica," pp. 16-17; and Alfred Holmes White, "Chemical Engineering education in the United States," *Transactions*, v. 21 (1928) pp. 55-59, 64. See also Olaf A. Hougen, "Seven Decades of Chemical Engineering," *CEP*, v. 73 (January, 1977) pp. 89-104.
67. For evidence of widespread dissatisfaction with current practices in chemical engineering education and the need of educational institutions for guidance see, for example, *Transactions*, v. 3 (1910) p. 122; *Bulletin*, no. 3 (1910) p. 5, and no. 4 (1911) pp. 17-18. For the appointment of the Committee on Chemical Engineering Education see *Transactions*, v. 1 (1908) p. 33. For examples of discussions of various experiments in chemical engineering education and the education problem generally see: M. C. Whitaker, "The Training of Chemical Engineers which Meets the Requirements of Manufacturers," *Transactions*, v. 3 (1910) pp. 159-68; F. W. Atkinson, "Development of the Chemist as an Engineer," *ibid.*, pp. 153-57; F. W. Frerichs, "Report of the Committee on Chemical Engineering Education ...," *ibid.*, pp. 122-45; "Report of the Committee on Chemical Engineering Education," *ibid.*, v. 7 (1914) pp. 18-31; v. 9 (1916) pp. 23-86; and v. 10 (1917) pp. 13-32.
68. *Transactions*, v. 7 (1914) p. 18; *Bulletin*, no. 8 (1913) p. 12.
69. *Transactions*, v. 7 (1914) pp. 19-22; *Bulletin*, no. 8 (1913) pp. 13-15, 44.
70. Herreshoff, "Training of Technical Chemists," p. 571.
71. *Transactions*, v. 7 (1914) pp. 23-25; *Bulletin*, no. 8 (1913) pp. 17-18.
72. *Transactions*, v. 7 (1914) pp. 40-42; *Bulletin*, no. 10 (1914) pp. 9, 24-31; no. 14 (1917) p. 10; no. 17 (1918) pp. 16-17.
73. *Transactions*, v. 12, pt. 1 (1919) p. 199.
74. Charles R. Mann, *A Study of Engineering Education*, New York, 1918.
75. White, "Chemical Engineering Education," p. 59.
76. R. T. Haslam, "The School of Chemical Engineering Practice of the Massachusetts Institute of Technology," *Journal of Industrial and Engineering Chemistry*, v. 13 (1921) pp. 465-68. Little's definition of unit operations is on p. 466.
77. Guédon, "Progetto dell'ingegneria chimica," pp. 21-23.
78. *Bulletin*, no. 20 (1919) pp. 50-53.
79. *Transactions*, v. 13, pt. 1 (1920) pp. 423-24; *Bulletin*, no. 21 (1920) pp. 17-19.
80. *Bulletin*, no. 22 (1920) pp. 26-28.

81. *Ibid.*, no. 23 (1921) pp. 15-24; "Report of Committee on Chemical Engineering Education of the American Institute of Chemical Engineers, 1922," separate booklet, no. p., 1922. See also Council, *Minutes*, February 16, 1921, and May 11, 1921.
82. "Report of Committee on Chemical Engineering Education, 1922."
83. *Ibid.*
84. *Bulletin*, no. 24 (1921) pp. 30-35. For Wesson's comment see p. 31, for McKee's see p. 34.
85. "Chemical Engineering Education Conference," *Journal of Industrial and Engineering Chemistry*, v. 14 (1922) p. 659; Council, *Minutes*, May 17, 1922.
86. *Bulletin*, no. 25 (1922) pp. 26, 30-32.
87. *Ibid.*, no. 31 (1925) pp. 19-25.
88. *Ibid.*, pp. 27-40.
89. *Transactions*, v. 19 (1927) p. 227.
90. See, for instance, H. A. Curtis, "Local Administration of Chemical Engineering Courses," *Transactions*, v. 17, pt. 1 (1925) pp. 243-62 (including discussion).
91. Council, *Minutes*, March 16, 1927.
92. *Transactions*, v. 27 (1931) pp. 403, 410.
93. White, "Chemical Engineering Education," p. 65. The concept of "unit processes" as something distinct from "unit operations" did not fully emerge prior to 1930. Before 1930 the two terms were sometimes, as here, used interchangeably. The context of White's comments mades it clear that he was referring to Little's "unit operations," not what later came to be called "unit processes."
94. Wisely, *American Civil Engineer*, pp. 80-81.
95. On the belief that schools would not allow professional societies to dictate the structure of their programs see *High Lights*, p. 17 (citing remarks by Bogert at the meeting of the Committee of Fifty), and Wisely, *American Civil Engineer*, p. 81.
96. *Transactions*, v. 32 (1936) pp. 570-571.
97. H. C. Parmelee, "Committee on Chemical Engineering Education," *Transactions*, v. 28 (1932) p. 322.
98. For instance, the Society for the Promotion of Engineering Education used chemical engineering as one of its five basic classifications of engineering in its *Report on the Investigation of Engineering Education, 1923-1929*, 2 vols., Pittsburgh, 1930-34.
99. Layton, *Revolt of the Engineers*, p. 4; Robert Perrucci and Joel E. Gerstl, *Profession Without Community: Engineers in American Society*, New York, 1969, pp. 10-12.
100. *High Lights*, p. 16.
101. Olsen, "Origin and Early Growth," p. 308.
102. Alfred Holmes White, "The Evolution of the Institute: The Adolescent Period," p. 6 (unpublished 16-page manuscript in AIChE files). White, 1929-30 AIChE President and an important figure in the early history of AIChE, wrote this manuscript c1950. It covers the period 1908 to c1932.
103. *Transactions*, v. 1 (1908) pp. 3, 21.
104. Samuel P. Sadtler, "Some Thoughts on the Organization of the 'American Institute of Chemical Engineers'," *ibid.*, pp. 42-43.
105. Frederick C. Zeisberg, "Committee on Publications," *Transactions*, v. 28 (1932) p. 319.
106. *Ibid.*, pp. 319-20. For the early AIChE's publications program see *High Lights*, pp. 41-45.
107. *Bulletin*, no. 33 (1926), p. 28.
108. Council, *Minutes*, May 10, 1929, and *Bulletin*, no. 40 (1929) pp. 35-47, for discussion of the publications program.
109. White, "Adolescent Period," p. 8.
110. *High Lights*, pp. 112-13, for early AIChE meetings policy.
111. Wisely, *American Civil Engineer*, pp. 130-31.
112. For the early history of engineering codes of ethics see Hunter Hughes, "The Search for a Single Code," *Consulting Engineer*, v. 15 (July, 1960) pp. 112-21, and William H. Wisely, "The

Influence of Engineering Societies on Professionalism and Ethics," in American Society of Civil Engineers, *Ethics, Professionalism, and Maintaining Competence,* New York, 1977, pp. 51-62.
113. *Transactions,* v. 4 (1911) pp. 50, 61.
114. *Ibid.,* v. 5 (1912) pp. 225-28; *Bulletin,* no. 5 (1912) pp. 23-25, 57-58.
115. *Bulletin,* no. 6 (1912) p. 24.
116. Layton, *Revolt of the Engineers,* pp. 84-85. The AIChE code was heavily influenced by the AIEE code; see, for instance, *Bulletin,* no. 5 (1912) p. 51. G. W. Thompson, "Committee on Ethics," *Transactions,* v. 28 (1932) p. 314, noted that because "other engineering societies had adopted codes of ethics . . . it seemed inevitable" that the AIChE should also.
117. *Bulletin,* no. 21 (1920) p. 15.
118. McKenna, "Justification," pp. 9-10; see also Sadtler, "Thoughts," p. 41.
119. *Transactions,* v. 1 (1908) p. 21.
120. Sinclair, *ASME,* pp. 41-42.
121. *Bulletin,* no. 6 (1912) pp. 18-20; no. 7 (1913) pp. 20-21; *High Lights,* pp. 88-89.
122. *High Lights,* pp. 87, 170.
123. *Bulletin,* no. 18 (1919) pp. 41-43; *High Lights,* pp. 87-88.
124. *High Lights,* p. 87.
125. *Bulletin,* no. 32 (1925) p. 38; *High Lights,* pp. 89-90.
126. *Bulletin,* no. 9 (1914) p. 12; no. 19 (1919) p. 31; no. 20 (1919) pp. 21-27; no. 23 (1921) pp. 44-49; no. 24 (1921) pp. 82-83.
127. "Symposium on the U.S. Patent System," *Transactions,* v. 4 (1911) pp. 417-505; William M. Grosvenor, "Report of Committee on Patents," *Transactions,* v. 4 (1911) pp. 503-04.
128. *Bulletin,* no. 6 (1912) pp. 18-20; no. 7 (1913) pp. 20-21.
129. *Ibid.,* no. 11 (1914) pp. 12-13.
130. *Ibid.,* no. 20 (1919) p. 38.
131. *Ibid.,* no. 26 (1922) pp. 34-35.
132. *High Lights,* p. 89.
133. *Ibid.,* p. 90; *Bulletin,* no. 33 (1926) pp. 40-41.
134. *Bulletin,* no. 40 (1929) pp. 35-37; Council, *Minutes,* April 12, 1929; June 18, 1929; September 9, 1929; December 1, 1929; and January 17, 1930; and Frederic J. LeMaistre, "Recent Growth and Development of the Institute," *Transactions,* v. 28 (1932) p. 325.
135. Kirkpatrick, ed., *Twenty-Five Years of Chemical Engineering Progress,* p. vii.
136. *Bulletin,* no. 45 (1932) p. 49.

CHAPTER **2**

Coming of Age, c1930–c1958

THE BREAK WITH ELITISM

When the AIChE established its first official headquarters in Philadelphia in 1930 it was still, in many senses, an exclusive club, representing not the profession as a whole, but an elite of the profession. Of its 872 members, 763 or 88% were "Active" or full members, 30 years of age or older, with five or more years in responsible charge in the chemical engineering field. What proportion of the entire chemical engineering profession were members of the AIChE in 1930 is difficult to determine, but it is almost certain that it was small, for the undergraduate enrollment in chemical engineering in American universities in 1930–31 was more than ten times greater (9154) than the total AIChE membership.[1]

As early as 1919 A. D. Little, then President of the AIChE, had called for the Institute to begin expanding its membership, pointing out that the "rigid procedures" used in admitting members excluded eminent college faculty and made some qualified chemical engineers reluctant to apply for fear of rejection. "I do not believe," he declared, "that the American Institute of Chemical Engineers will ever be accepted as the organization qualified to act for the Chemical Engineers of this country, until its membership is very much larger."[2] In 1920 Vice President Henry Howard (President, 1922–23) repeated Little's call for the Institute to expand its membership: "We should all agree," he stated, "that the Institute ought to be truly representative of American chemical engineers." "We should use our best efforts to obtain as members of the Institute all of the properly qualified chemical engineers."[3]

But Little and Howard were in the minority. An amendment proposed by Little to permit Council and the Membership Committee to waive normal requirements to permit the admission of candidates of unusual merit or achievement was defeated.[4] Among those opposing a broader membership base were some of the most prominent figures in the early organization. John Olsen, the Institute's Secretary, defended the exclusive nature of the AIChE.[5] David Wesson, 1920–21 AIChE President, felt that 600 was a desirable size for the Institute. He argued that if the organization became too large it would "defeat the object of sociability among its members."[6] H. K. Moore, a Vice President from 1920 to 1925 and President in 1926, argued that even 600 was too large. He maintained that the existing size (around 300 in 1920) was sufficient and praised the club-like atmosphere of the Institute.[7]

Although the proponents of a restricted society continued to dominate in the early 1920s, several factors were working in favor of those who wished to broaden AIChE's membership. One of the major factors was the continued growth of the chemical industry in America and, consequently, of the chemical engineering profession.

The rapid pre-1914 growth rate of the American chemical industry had been further accelerated beginning in 1914–15 with the onset of World War I and the suspension of German chemical imports into the United States. The industry had continued to expand at a rapid pace until 1920. After experiencing a severe slump from mid 1920 to the end of 1921, it had resumed rapid growth, encouraged by the tariff protection of the Fordney-McCumber Act of 1922 and by numerous mergers and increased concentration.[8] Between 1913 and 1927, according to Fritz Haber's estimates, the value of American chemical products increased from 34% of the world's total to 43%.[9]

The onset of the Great Depression in 1929–30 slowed down the growth of the American chemical industry. But the impact of the depression on the chemical industry was not nearly as serious as it was on most other industries. Chemical prices, for instance, declined only 7.3% during the sharp price deflection of 1930, while the price of "all commodities" dropped 23.6%.[10] Haynes' survey of 45 leading chemical manufacturers for the period between 1929 and 1937 indicated that 38 continued to pay dividends regularly, that 35 paid larger dividends in 1937 than 1929, and that the average earnings of the group had increased by 28%.[11] Chemical output passed the record year of 1929 in 1937; industry as a whole did not achieve a similar performance until 1941, under the stimulus of World War II.[12] Further, in the 1930s the chemical industry did not cut back on research. The number of research workers in the chemical industry grew from 3451 in 1927 to 9467 in 1938.[13] While not exactly a "depression-proof" industry, the chemical industry was certainly more depression proof than most other industries in the 1930s.

In the 1940s, under the stimulus of World War II, the chemical industry resumed the rapid growth rate it had enjoyed in the 1920s and continued in this pattern through the 1950s and 1960s. The "value added" to chemical and allied products (basic chemicals, chemicals used in further manufacturing, and such things as drugs, soaps, fertilizers, and explosives, but excluding such industries as petroleum, coal, rubber, and plastics) increased from $1.74 billion in 1929 to only $1.82 billion in 1939. But the figure had reached $7.24 billion in 1950; $14.42 billion in 1960; and $27.93 billion in 1970.[14] In 1929 the chemical and allied products industries employed 382,000; in 1970 the figure was 881,000.[15]

Both the rapid growth of the chemical industry in properous times and its relatively good performance during the depression stimulated the growth of the chemical engineering profession. For instance, Albert Newman (Director, 1940–42; President, 1948) observed in 1938 that because the chemical industries had been affected by the depression only "to a relatively small degree," chemical engineering graduates had continued to find employment "fairly readily."[16] This, he added, had caused a rapid influx of students into the field. In 1927–28 chemical engineering students totalled 5702 and formed 11.4% of the total number of undergraduates enrolled in the four leading engineering fields: civil, mechanical, electrical, and chemical engineering. In 1936–37 there were 12,556 chemical engineering undergraduates, and they formed 24.5% of the enrollment in these four fields. In addition, in 1937–38 chemical engineering students formed the largest single group among the 2099 engineering students studying for the master's degree and the 362 enrolled for the doctorate.[17]

Although the return of prosperity diminished the proportion of engineering students becoming chemical engineers, reducing it to the more usual 9 to 10% by the late 1940s,[18] the profession continued to grow in absolute terms. Between 1940 and 1950 the number of chemical engineers reported in census returns increased over 160%, between 1950 and 1960 the increase was 23%, and between 1960 and 1970 it was 26%.[19]

The steady growth in the number of chemical engineers, stimulated by the near-steady growth of the American chemical industry, presented the Institute with a difficult choice. It could continue to maintain very high entrance requirements and sharply limit its membership, or it could actively seek to expand its membership base. If it followed the first alternative, it would continue to provide prestige to those admitted to its ranks and enjoy the benefits of smallness. But it risked losing its claim to being spokesman for the chemical engineering profession if the 90% or more of the profession not permitted to join the AIChE formed a rival group or threw their support behind the American Chemical Society. On the other hand, if it enlarged its membership, the club-like atmosphere valued by many of the Institute's early members would be lost and the prestige of being an AIChE member would be significantly reduced. But by broadening the Institute's membership and appeal, the organization could sustain its claim to represent the chemical engineering profession.

A Change in Membership Policy

The first sharp break with the restrictive membership policies that had dominated the early history of the Institute came in 1930-31. The stimulus was provided by the American Society of Mechanical Engineers' plan to create a Process Industry Division which would include, among others, chemical engineers involved in the operation and maintenance of chemical processing plants. Worried about the potential impact of this division, concerned Institute members introduced a proposal to amend the AIChE's membership requirements at the June, 1930, Detroit meeting.[20] In the 1908 constitution "Active" members were required to be 30 years old and "proficient in *chemistry and in some branch of engineering* as applied to chemical problems." In 1928 this had been modified to read "proficient in *chemistry and* the design, construction or operation of a plant in which matter undergoes a change of state or composition." The wording proposed at the June, 1930, meeting called for Active members to be "proficient in *chemical engineering and* experienced in the design, construction, maintenance or operation of a plant in which matter undergoes a change of state or composition."

Those favoring the proposed changes argued that many qualified chemical engineers were not applying for membership in AIChE because they regarded it as a society of chemists instead of engineers and because they feared that the requirement of proficiency in chemistry implied expertise in analytical chemistry and laboratory manipulation. Those opposing feared that the change would open the doors of the Institute to men with inferior qualifications and inadequate chemical training.[21]

When submitted to the entire membership by mail ballot in the summer of 1930 the amendment passed by a 385 to 106 margin, barely over the 75% needed to approve constitutional changes.[22] But the issue was not so easily resolved. On October 17, 1930, A. C. Langmuir, a charter member of the Institute and long active in its affairs, asked Council that the amendment be resubmitted because proper time for consideration and debate had not been provided. There was some basis for his appeal, but Council rejected the request by a nine to three vote.[23] At the next meeting of Council, on November 7, 1930, Langmuir submitted a formal petition to amend the constitution by requiring Active members to be "proficient in *chemistry and chemical engineering*," instead of simply "proficient in chemical engineering." He also announced the tentative results of a straw ballot he had sent members on the issue—326 had favored his proposal, only 30 had opposed.[24]

Langmuir's proposal was considered at the New Orleans meeting of the Institute in December, 1930. The debate, according to one observer, was "prolonged and sometimes acrimonious."[25] Supporters argued that it was impossible to practice chemical engineering successfully without proficiency in chemistry and that the membership requirements passed in 1930 would ultimately damage the prestige and influence of AIChE. In opposition, J. M. Weiss, a prominent consulting chemical engineer with numerous patents, contended that chemical engineering as a profession had "grown up": "It needs no explanation and no addition. It is a sufficient term to define properly what we are now doing and what we will do." Crosby Field (Director, 1931–32) feared that the proposed amendment would restrict the Institute's future growth and warned that the AIChE's hold on the profession was tenuous. On the one side, he noted, the ACS with its 20,000 members was entering the field through its industrial division. On the other side the ASME was pushing into the field through its newly-created Process Division. Expansion was necessary. Convinced by this rhetoric, those present voted 45 to 17 against Langmuir's proposal to require proficiency in chemistry *and* chemical engineering as a requirement for membership.[26]

After the New Orleans meeting, Council also reviewed the proposed amendment and also voted (by a nine to two margin) to go on record favoring its defeat. A. E. Marshall (President, 1934–35) stated Council's position:

> We believe that Chemical Engineering, as a profession, has become definitely established and that it is just as unnecessary and undignified to qualify Chemical Engineering in terms of Chemistry, as it would be to qualify Mechanical Engineering in terms of Physics.[27]

Langmuir's proposal failed by the narrowest of margins when submitted to the membership. Of those returning the mail ballot 362 or 69% favored requiring proficiency in "chemistry *and* chemical engineering," 163 opposed inserting the requirement of proficiency in chemistry.[28] The amendment had fallen barely 30 votes short of the 75% majority required for passage.

Several questions were at issue in the debate over this amendment. In one sense it was a dispute between those who felt chemical engineering had fully obtained the status of an independent discipline and wished to completely leave the embrace of the discipline's parent, chemistry, and those who wished the new discipline to retain very close ties with its parent. But at another level, the dispute was between those advocating sharply restricted membership standards and those favoring a broader membership policy. In general, those who supported Langmuir were the older members of AIChE, those who had long dominated the organization and wished to keep it an elite group. Those who opposed him were largely the younger members of the Institute who wished to broaden the Institute's membership and appeal.[29]

The constitutional debates and balloting of 1930–31 (balloting on Langmuir's amendment occurred in early 1931) were a watershed. After 1930–31 those favoring a break with very restrictive membership qualifications were increasingly ascendant. Membership statistics do not immediately reflect this change. The AIChE growth rate for the period 1930-34 was even slightly lower than it had been between 1926 and 1930. But this lower growth rate probably reflects the onset of the depression more than anything else. The changing of the guard is better reflected in the rising proportion of members in the Institute's lower member grades. Between 1926 and 1929 the proportion of "Junior" members in the AIChE declined from 16% to 11%. The Junior grade was in so little demand that the Institute abolished it in 1929 and created an Associate grade with somewhat different membership requirements. Between 1930 and 1935, however, the proportion of non-Active members (i.e., Associates and—after the recreation of the Junior grade in 1933—Associate and Junior members) grew steadily, rising from 12% in 1930 to 21% by 1935.

ORGANIZATIONAL CHANGES

Thus, over significant opposition, the American Institute of Chemical Engineers embarked on the path of broadening its membership in the early 1930s, hoping to retain and consolidate its claim to represent the chemical engineering profession. Because of this, the Institute's second quarter century stands in stark contrast to its first. In its first quarter century much of its energies had been absorbed in external matters, particularly the regulation of engineering education. Internal organization had attracted little attention. In contrast, the Institute's energies in its second quarter century were largely absorbed by the gradual creation of an organizational structure and programs to attract and to hold a wider membership. Some of the important organizational changes and programs undertaken by AIChE between 1930 and 1958 were:
- establishment of a permanent headquarters with full-time staff;
- initiation of more aggressive membership campaigns;
- modification of membership requirements to permit broader participation in the organization;
- enlargement of the student chapter program;
- enlargement of the local section structure and creation of a technical division structure; and
- initiation of broader and stronger programs in the technical field, especially a strengthened publication program.

Headquarters

One of the most obvious requirements for any sustained attempt to expand the AIChE from a small, exclusive organization into an organization representing the entire chemical engineering profession was a permanent headquarters with a full-time staff and a full-time administrator. For almost 20 years (1908-26) John Olsen's desk had served as the Institute's offices, and Olsen had attempted to handle AIChE business in his spare time, compensated only by the satisfaction the job gave him and by a modest expense account and honorarium. From 1927 to 1929 Olsen's desk was replaced by H. C. Parmelee's. But already by the late 1920s the volume of the Institute's business was becoming too large for spare-time efforts.

The possibility of establishing a permanent headquarters office was raised by Council in December, 1927, but rejected on financial grounds.[30] The issue was revived in 1929. In

April, Secretary H. C. Parmelee informed Council that the rapidly growing duties of the office were making it imperative to establish a more permanent and adequately paid secretarial organization.[31] His position was reinforced by the report of a special committee appointed to review the growing deficiencies of the Institute's publications program. This committee blamed some of the shortcomings of the *Transactions* on the long lag time between when papers were presented and when they appeared in print. A paid Executive Secretary, the Publications Committee believed, would reduce the problem. F. C. Zeisberg, the Publications Committee's chairman, considered the appointment of such a secretary an "absolute necessity"[32] and observed that a questionnaire his committee had submitted to AIChE members had yielded a 304 to 7 margin in favor of employment of a full-time secretary on salary.[33]

In the fall of 1929 the AIChE began to seek an Executive Secretary.[34] In January, 1930, after a review of the efforts made to obtain the services of a man willing to devote at least half time to the management of the Institute's affairs, Council offered the position to Frederic LeMaistre, a Canadian-born and educated (McGill, 1904) chemist and chemical engineer, a former employee of Du Pont (1906-20), and a private consultant (since 1920). He accepted the position on a half-time basis, continuing his regular consulting practice.[35] Because this practice was located in Philadelphia, the Institute established its first formal office in that city.

By 1937, however, the growth of the AIChE (its membership was then well over 1000) led to considerable work, more than could be handled on a part-time basis. LeMaistre urged the Institute to appoint a full-time Executive Secretary to replace him so that the business of the organization could be carried out more effectively.[36]

Council accepted his advice and appointed Stephen L. Tyler as its first *full-time* Executive Secretary. Tyler, a 1911 Yale graduate in chemistry with graduate work in metallurgy, had been employed for some years as chemical engineer and U. S. sales manager for Thermal Syndicate, Ltd., a British firm that manufactured and sold fused quartz and other ceramic materials. He had, in addition, invented and patented the "Tyler Absorber," widely used for some years in the hydrochloric acid industry, and served as chairman of the New York section of AIChE.

At the same time, the society's offices were moved to the Engineering Societies Building in New York, where most of the other major American engineering societies had their offices. Tyler's staff grew from a single employee in 1937 to 24 by 1947 and to 49 by 1954, the year he retired.

The impact on Institute growth of hiring a full-time executive of Tyler's caliber and employing a headquarters staff is clear. There was a definite increase in the rate of growth beginning in 1937, the year of his initial appointment. Between 1932 and 1936 annual growth had averaged 7.2%. In the first five years following Tyler's appointment growth averaged 15%.

Membership Campaigns

Establishment of a full-time headquarters and staff, however, was just one of the ways the Institute broadened and improved its services and expanded its membership rolls in its second quarter century. Equally important was the initiation of more aggressive membership campaigns. In the early years the Institute had briefly considered a limited campaign to increase membership, a campaign which involved approaching a carefully screened list of qualified and desirable candidates. But in the early years there was sufficient disagreement over the advisability of doing even this, that the plan was dropped.[37] In 1911 the secretary reported: "Most of the new members are obtained by the personal solicitation of our present membership."[38] This approach—personal solicitation of individual prospective members—seems to have been the dominant method for securing new members throughout the 1910s and 1920s.

The threat of ASME entrance into the domain of chemical engineering through a Process Division and the modification of membership requirements in 1930 to stress proficiency in chemical engineering instead of chemistry signaled the beginning of a more aggressive policy. This policy was further encouraged by the Great Depression. The early depression adversely affected AIChE finances. Sales of the *Transactions,* for instance, dropped considerably in 1930 and 1931.[39] Much of the Institute's reserves had been invested in bonds which fell sharply in value and ceased paying dividends.[40] As a result, the 1931 budget ran at a deficit.[41] In 1932 the Institute appointed a Committee on Revision of Expenditures, reduced the salaries of its employees, moved to smaller offices, resigned from the American Engineering Council (membership required dues), and published only a single volume of *Transactions.*[42] In January, 1932, in the midst of this financial crisis, Council considered ways of increasing revenues. It reviewed three options: increasing membership, increasing dues, and creating an endowment fund. Increasing dues seemed inadvisable; creating an endowment fund seemed unlikely to succeed. But increasing membership seemed to hold promise.[43] In October, 1932, Council voted to initiate in 1933 a far more aggressive membership campaign than any previously conducted. Council requested the Executive Secretary to write to the professors in charge of chemical engineering at each accredited university and to the counselors of student chemical engineering organizations, requesting them to present information on AIChE membership to their graduating classes.[44]

This approach was replaced with a more moderate program several years later. Beginning in 1938 the Institute's Membership Advisory Committee solicited lists from chemical engineering professors of graduates who had been out of school for three to four years and who had "demonstrated their ability to give credible service in Chemical Engineering." Such lists usually constituted around 25% of a graduating class. Men on these lists were then canvassed to determine if they were interested in membership.[45]

Although the membership campaigns conducted in the 1930s were limited primarily to chemical engineers just beginning their careers and did not attempt to reach academic, mid-career, or upper management engineers, they were a far more aggressive approach to expanding membership than previous practice and represented a sharp break with earlier philosophy. Since these membership campaigns were primarily aimed at young chemical engineers who would enter as Junior members, the success of the campaigns can be judged, in part, from the growing proportion of such members in AIChE. In 1929 only 10.6% were Junior members. In 1940 the proportion had risen to 33%. This growing contingent caused some concern among more conservative members, and in 1938 John Olsen, who had been conducting the membership campaigns, was asked to refrain for a time from soliciting Junior members and to seek, instead, new Active members.[46] But the trend proved irreversible. In 1944, for the first time in its history, the organization contained more Junior than Active members.

In 1950 the Institute further broadened its campaign for new members by creating a new Membership Committee. This new committee, initially chaired by C. G. Kirkbride

(President, 1954), was asked by Council to take a much more aggressive attitude in securing new members. It was, in effect, charged with trying to sell the Institute to all practicing chemical engineers. This was, as Van Antwerpen and Fourdrinier noted in their fifty-year history of the AIChE, a "rather radical departure" for the Institute. It had never before had a committee which so actively encouraged and promoted membership growth, which tried to sell the Institute instead of just making information about it available. Among the new committee's programs were sponsorship of competition between local sections for new members and more sustained efforts to attract new graduates by making it easier for students to join on graduation. Under John McKetta's (President, 1962) chairmanship the program to attract new members was further expanded by encouraging each local section to initiate membership drives of its own.[47]

To strengthen the ties between the new, younger members and the Institute and to make the organization more attractive to them, Council in 1943 authorized a Committee on Junior Activities. This new committee was to provide a means for younger members to furnish input into the national organization and to attempt to place more of them on Institute committees. In the year following its creation, the number of Junior members on national committees increased from 6 to 31.[48] In 1945 the committee was responsible for creating a special award (formerly the Junior, now the Allan P. Colburn, Award), limited to members under 35.[49] It was also active in promoting the constitutional changes (reviewed below) which altered the name of the Junior grade to Associate and increased the privileges of Associate membership.

Membership Requirements

In conjunction with its attempt to broaden AIChE's membership base through more active recruiting efforts, Council also sought to broaden the base through modifications in constitutional requirements for membership. The first major change in this direction, as noted earlier, was taken in 1930 when proficiency in chemical engineering replaced proficiency in chemistry in the requirements for Active membership. In 1943 membership requirements were modified again to permit chemical engineers involved in industrial research and in teaching to enter the Institute as Active members. In 1950 the Institute approved the automatic transfer of Junior members to Associate membership at age 35. Previously members failing to apply for either Active or Associate membership at 35 were dropped from the society. Finally, in 1954 the somewhat degrading name "Junior" was abolished. The Junior members became Associate members; the former Associate members became Affiliate members. At the same time the new Associate members were given the right to vote in all save constitutional elections and to nominate officers, privileges previously restricted to the Active members.[50]

Student Chapters

The bulk of the new membership in most professional engineering societies comes from new graduates. Thus the expansion of and improvements made in the AIChE's student chapter program between 1930 and 1958 were as critical to the growth of the organization as the improvements made in administration, as its general membership drives, and as its modified membership requirements. The student chapter program put it into contact with chemical engineering students in their formative years, provided a means for funneling new graduates into the organization, and generally increased the likelihood that they would join AIChE on graduation.

Because of the Institute's intense early concern with education, it is not surprising that the question of student chapters came up relatively early. In 1913 John Olsen, the Institute's secretary, reported receiving two applications for student societies. But the AIChE took no action.[51] Several explanations for this inactivity are possible. First, in its early years it was a relatively small and exclusive organization, and many of its members wished to keep it that way. Second, consciously or unconsciously the Institute may have desired to avoid, this early in its history, direct competition with the ACS for the membership of new graduates in chemistry and chemical engineering. Finally, and most likely, in 1913 very few schools had an Active member of the Institute on their teaching staffs to sponsor such organizations. It was certainly for the latter reason that Council rejected the idea of student chapters when it was brought up again in 1917.[52] Other engineering societies in the meantime had begun to establish them. ASME, for instance, established its first student chapter in 1908, the year AIChE was organized; ASCE authorized them in 1919.[53]

In 1921 Alfred Holmes White, an academic engineer elected to Council (and later President, 1929–30), reintroduced the idea of creating local student chapters, in spite of being warned that "sentiment in the Council was strongly against such expansion."[54] Council, however, approved the idea in June, 1921, turning the matter over to a committee to draft the appropriate amendments to the constitution. The amendments were approved at the December, 1921, meeting of the Institute and by the entire membership by letter ballot in early 1922.[55] At the December, 1922, meeting AIChE authorized its first student chapter, organized by White, at the University of Michigan.[56]

Although the Institute had 15 student chapters by 1929, its links to these chapters were few. In 1928, for instance, Professor H. L. Olin of the University of Iowa complained that the Institute paid little or no attention to its chapters and provided no encouragement to keep the chapters alive and active.[57] In response Council created in 1928 a temporary and in 1929 a permanent Committee on Student Chapters. The early efforts of this committee focused on attempts to secure at least two visits to each chapter annually by Institute members.[58] In 1932, however, the committee introduced the Student Problem Contest, an annual competition open to all undergraduate members of AIChE student chapters. The contest problem, similar to problems that a young engineer might face on the job, was prepared by a subcommittee of engineers in industry and distributed by student chapter counselors. Solutions were judged by the subcommittee, with first, second, and third prize winners being honored at an awards banquet at the Annual Meeting of the Institute and their solutions published.[59] This contest quickly became a permanent, annual Institute activity.

Further ties to student chapters were created in subsequent years. The *Student Chapter Bulletin*, begun in 1937 and discontinued in 1946 (but revived later), contained chapter news and relevant Institute news and propaganda. It was replaced in 1948 by a short newsletter, *Student Chapter News*. But this, too, had been discontinued by the late 1950s, to be replaced by the *Student Member Bulletin*. In 1954 AIChE amended its constitution to create a class of Student members. Additional modifications to the constitution and bylaws in the late 1950s made Student member and member of a student chapter practically synonymous and provided easy methods for the transfer of Student members to the Associate (formerly Junior) grade upon graduation.[60]

The success of the early student chapter program is indicated by its growth. The first AIChE student chapter was formed in 1922. There were 15 chapters by 1929. By 1938, reflecting both greater AIChE interest and the prosperity enjoyed by chemical engineering programs during the depression, there were 64 student chapters with 4875 members. These included one-third of the total number of chemical engineering undergraduates and 75% of all eligible chemical engineering seniors.[61] At the Institute's 50th anniversary in 1958 it had 108 student chapters with a total membership of 7035. Student members composed about 42% of all chemical engineering undergraduates and about 74% of eligible seniors.[62]

Local Sections

Local sections—local organizations of national society members—generally served several purposes in American engineering societies during the twentieth century. First, they helped offset the geographical imbalance created by the necessity of having to hold national meetings only in the few urban areas with sufficient lodging and meeting facilities to accomodate them. Second, because they were smaller and met more regularly, they were better able to encourage the formation of closer professional and personal relationships among members. Third, they provided means for younger engineers to demonstrate their leadership and organizational talents and to participate in professional society affairs. In almost every American engineering society, non-supervisory and mid-career engineers played a much more significant role in governing local sections than in governance at the national level. Finally, local sections were better able to adjust programs to local member needs.

The older and larger societies, like ASCE, had, of course, encountered the problems of geographical imbalance and local needs earlier than AIChE. The American Society of Civil Engineers had considered the creation of local sections as early as 1890-91, had adopted a set of model bylaws to provide for the creation of local associations of ASCE members in 1905, and had formally amended its constitution to give official status to local associations in 1921.[63] The ASME had approved rules for local branches in 1905 and approved the first of its local sections between 1909 and 1912.[64] The AIME established its first local sections in 1911; AIEE in 1902; the Institute of Radio Engineers (IRE) in the year of its founding, 1914.[65]

The possibility of forming local sections of AIChE was briefly broached as early as 1911, but there was general agreement that it was not large enough or strong enough to organize local sections.[66] In 1921 Council began to seriously consider the problem again, and the possibility of organizing local sections was discussed at the December, 1921, Baltimore meeting. Opinion there was favorable. Both Cleveland and Detroit already had informal organizations of AIChE members which had emerged without any reference to the parent organization.[67] Thus Council in January, 1922, drafted rules for organizing local sections. These required local sections to have at least ten members, meet at least four times a year, have their officers approved by Council, and limit their meetings to hearing and discussing papers and to informal fellowship.[68] One section, at Detroit, was organized under these rules in mid 1922, but it did not survive the depression.[69] The first permanent local section of AIChE was the Chicago section, founded in 1924-25.[70]

The Institute's local section structure did not grow rapidly at first. The 1929 *Directory* of the AIChE lists only two—Detroit and Chicago. In the early 1930s, however, Council began to devote additional attention to the creation of local sections, feeling that more were needed. Secretary LeMaistre made a comparative study of the local section organizations of other societies for Council in 1932,[71] and Council created, at about the same time, a Committee on Local Sections.[72] In the early 1930s the Council, hoping to encourage local groups, also loosened the requirements earlier imposed on them, granting wide privileges in the way local associations conducted meetings and permitting them to invite whom they wished to meetings.[73] The growth rate accelerated. By 1940 AIChE had 12 local sections, by 1945 it had 19, by 1950 it had 37, and by 1957 it had 62. Membership in local sections in 1957 was 10,869 (7927 were AIChE members).[74]

AIChE's local sections, like those of most engineering societies, had a great deal of autonomy and independence, especially after the early 1930s. They elected their own officers, established their own dues structure, and put together their own programs. The national organization, at first, reserved only the right to formally approve the constitution and bylaws of the local sections and to prohibit local sections from making financial commitments which would obligate the national organization and from taking political stands in the Institute's name.

In the 1940s and 1950s the AIChE began to strengthen its loose ties to its local sections. By the late 1950s national had established a set of model bylaws to guide local sections and had prepared a manual for local section officers. Other services included setting geographical boundaries of local sections, providing mailing lists of eligible national members in the area, distributing a local section newsletter, and publicizing local section activities in the Institute's *Bulletin* (prior to 1947) and in *Chemical Engineering Progress* (*CEP*) (after 1947). The Institute's most important contribution to the health of its local sections, however, was its Speaker's Bureau, created in 1948 to improve local section programming by informing them of good speakers willing to make presentations for travel expenses only.[75]

Local sections in the 1940s and 1950s occasionally asked for more services from the national organization, in particular for some form of dues rebate. But the Institute's policy through its second quarter century was to insist that they pay their own way with national providing all the help it could short of financial aid.[76]

Divisions

The creation of a local section structure helped provide the AIChE with geographical balance and local connections which broadened the appeal of membership. The creation of a technical division structure in the 1950s to parallel this local section structure, helped provide it with a technical balance which increased its appeal to specialists.

Generally, as the broad engineering disciplines matured and grew more complex in the late nineteenth and the twentieth centuries, the proportion of specialists within these disciplines increased and that of generalists decreased. This evolution created problems for societies attempting to unify and represent broad engineering disciplines because they had been founded and long governed largely by engineers with generalist inclinations. To prevent the formation of splinter societies by discontented specialists, most engineering societies adjusted to the demands of specialization by creating technical divisions, essentially sub-societies with considerable autonomy in internal organization and programming.

Because most of the other major engineering disciplines, for instance civil and mechanical engineering, were older than chemical engineering, the societies which represented them encountered the need to modify internal organization to accomodate specialization much earlier than the AIChE. For example, the ASCE considered creation of technical divisions in 1873 but rejected them. As a result civil engineers interested in specialties like railroad engineering and water works engineering split off and formed splinter societies late in the 1800s. To prevent further splintering the ASCE created a formal divisional structure in 1922.[77] The ASME also approached the creation of technical divisions with considerable reluctance. Hence mechanical engineers concerned with heating and ventilation split off in 1894 to form their own society; mechanical engineers interested in refrigeration did the same in 1904; and mechanical engineers associated with the automobile industry formed the Society of Automotive Engineers in 1905. ASME belatedly reponded by creating its first technical division—a Gas Power Section—in 1907.[78]

In the first three decades of its existence, the AIChE was small enough and the vitality of chemical engineering as a legitimate discipline so tenuous that subdivision was almost unthinkable. Often the Institute's Papers Committee had difficulty securing enough high quality papers for the Institute's national meetings. In such circumstances it is unlikely that a specialty division could have generated sufficient literature or sufficient enthusiasm to support itself.

By the 1940s, however, the chemical engineering profession was large enough and complex enough to begin fostering self-sufficient specializations willing to form splinter societies if their needs were not met. The formation of the Society of Plastics Engineers in 1941 was a portent of things to come if the AIChE did not respond to the needs of its specialists.

Technical divisions ultimately emerged in AIChE in much the same manner as they had emerged in other engineering societies. They began as an extension of the Institute's committee structure. A special area within the Program Committee would be formed for a particular specialization and given considerable freedom in organizing programs for meetings and soliciting papers. As this committee proved viable, it would be upgraded to divisional status with its own governmental structure and autonomy in electing officers, charging dues, and managing its own affairs within limits.

The first technical division emerged in 1954. In 1947 Council approved the formation of a Nuclear Engineering Committee to encourage work in the nascent area of atomic power through the promotion of research and programming. In 1954 this committee organized the first international conference on the peacetime use of atomic energy at Ann Arbor, Michigan. The engineers present at that conference organized AIChE's first division, Nuclear Engineering, adopting for this division bylaws permitting membership to anyone interested in the field of nuclear energy with no requirement of AIChE membership, in the hope of providing a home for all engineers and scientists in the nuclear field. Although most engineers joined this division, the scientists in attendance chose, instead, to form an independent society—the American Nuclear Society.

AIChE's Nuclear Engineering Division was active in programming during the 1950s, sponsoring, for example, six sessions on the nuclear fuel cycle at the 1957 Nuclear Congress in Philadelphia, a symposium on the role of industry in university nuclear engineering programs at the June, 1957, AIChE meeting, and another symposium at the April, 1958, meeting in Montreal.[79]

A second division—Heat Transfer—emerged in 1958 after several years of activity as a committee and was similarly active in programming. In 1957 and 1958, for instance, it cooperated with the ASME in two heat transfer conferences.[80] These annual Joint Heat Transfer Conferences have continued to date.

TECHNICAL PROGRAMS

The establishment of a permanent headquarters and paid staff; more aggressive membership campaigns; modifications to the sections of the constitution governing membership qualifications and privileges; and the creation of student chapters, local sections, and technical divisions can be regarded as the stones with which AIChE's expansion in the 1930s, 1940s, and 1950s was built. But the mortar which bound these stones together was the Institute's technical publications and meetings. In both of these traditional areas the Institute expanded its programs and broadened its appeal in its second quarter century. In addition, it expanded into new technical areas like standardization and research where its prior involvement had been either miniscule or non-existent.

Publications

Because the most regular contact the average member of a professional society has with the national organization is through its publication program, and because publications are often the most tangible benefit of membership, the creation of an appealing publication program was a vital element in AIChE's attempts to broaden its appeal in the period from 1930 to the middle of the 1950s.

For almost 30 years the AIChE's publications program was largely limited to the issuance of annual transactions. This policy, however, had led to problems by the late 1920s, as noted in the previous chapter. A special committee on publications policy, chaired by H. A. Curtis (Director, 1927–29, 1931–33), recommended in 1929 that the AIChE abandon its *Transactions* and replace them with an annual review of progress in the different unit operations areas; release papers to authors immediately after presentation at meetings so that they would be more willing to present good papers at these meetings; and hire a full-time Executive Secretary to solicit better papers and to expedite their publication. A questionnaire submitted to membership by the committee indicated broad support for all of these propositions. The vote to discontinue the *Transactions,* for instance, was 217 to 79.[81]

At the December, 1929, meeting of the Institute, publications were discussed at length. At that meeting some of the organization's most prominent members spoke against abandoning the existing publications program, pointing out that review-type journals already existed in the chemical engineering field. Others argued that the Institute could not afford simultaneously to hire an Executive Secretary and initiate a new journal of reviews. As a result, H. A. Curtis moved that the Institute concentrate first on hiring an Executive Secretary (in the hope that this would alleviate many of the problems the Institute was encountering in securing papers and publishing them expeditiously), continue its *Transactions,* and experiment with the publication of reviews within current publications. The motion passed.[82]

In reviewing the Institute's publications program in 1932 Fred C. Zeisberg (President, 1938) reported some improvements. He noted, for instance, that for the first time in recent years there were more papers submitted than could be accepted.[83] But A. G. Peterkin (Director, 1934–36), Chairman of the newly-formed Papers Committee, still reported a "decided reluctance" on the part of some "prominent members" to contribute papers to AIChE because of the lag in the publication of the annual transactions and because of a preference for the monthly publications of ACS.[84] The problem continued to plague the Institute through the 1930s. In 1936 Council reviewed with concern the ACS's sponsorship and publication of annual symposia in chemical engineering, feeling that they were an invasion of AIChE's territory. F. C. Zeisberg, then serving as the Institute's Vice President and a former chairman of its Publications Committee, argued that the roots of this problem lay in the dissatisfaction with publication policies and priorities of certain younger engineers interested in papers dealing with the fundamental principles of chemical engineering.[85]

The Institute took several steps in the 1930s to correct the deficiencies in its publications program and to lure its members away from ACS meetings and publications. In 1935, to speed up publication of papers presented at AIChE meetings, it began to issue the *Transactions* as paper-bound quarterlies, and in 1938 as paper-bound bimonthlies. The Papers Committee agreed to increase the proportion of papers devoted to basic principles, and Council invited those engineers who had organized the annual chemical engineering symposia through ACS to hold the symposia in conjunction with AIChE meetings.[86]

These steps were in the right direction but were clearly insufficient to meet the competition of ACS's publications program which was better financed and had broader coverage. In January, 1946, on the recommendation of George Granger Brown (President, 1944), then chairman of its Publications Committee, Council voted to take more aggressive steps to improve its publications program. It authorized the withdrawal of $55,000 from reserves to hire a full-time editor and begin publication by January 1, 1947, of a monthly journal which would carry, in addition to the papers presented before the Institute, Institute news and reports and general items and articles of interest to chemical engineers.[87]

Franklin J. Van Antwerpen, previously active in a number of Institute programs, was hired in 1946 to direct the new program. A future Executive Secretary of the Institute, he was a chemical engineering graduate of Newark College of Engineering (now the New Jersey Institute of Technology) and came to AIChE from the American Chemical Society, where he had served as managing editor of *Industrial and Engineering Chemistry*.

Under Van Antwerpen's direction, the Institute launched *Chemical Engineering Progress* (*CEP*) in 1947. This new monthly journal was a major break with AIChE tradition. The *Transactions* (which were slowly phased out in the late 1940s and early 1950s) had been composed almost entirely of formal papers bound in a conservative, scholarly 6 × 9 inch format. They had appealed mainly to serious scholars and research engineers. *CEP* used an 8.5 × 11 inch format. It was an illustrated magazine and carried industrial news and features as well as formal papers. Gradually, over the next decade, its contents and format were improved and modified to appeal to a steadily wider readership. The creation of a new flagship publication and the employment of a full-time professional editor naturally increased publication costs. But these were partially offset by permitting the new journal to carry advertising. This, too, was a break with tradition. The AIChE had previously relied entirely on membership dues for financial support of all its programs, including publications.[88]

The introduction of *CEP* did much to allay the publication woes of AIChE. The old *Transactions* had published an average of 37 papers per year in the last ten years of its existence. The capacity of *Chemical Engineering Progress* was, by 1950, about 95 papers per year.[89] Unfortunately, due to the growth of the AIChE generally and AIChE meetings in particular, even this was insufficient. By 1950 almost 150 papers were being presented annually at AIChE meetings.

To fill this gap the Institute introduced its *Symposium Series* and its *Monograph Series* in 1951. The *Symposium Series* volumes contained collections of papers on a single topic organized by volunteer editors. They were published at irregular intervals and were of interest primarily to specialists. By 1958 20 volumes had issued in this series. The *Monograph Series* consisted of book-length works of interest to chemical engineers. Its volumes were published at even more irregular intervals.[90]

Finally, around its 50th anniversary the AIChE began publishing several series of technical manuals. The *CEP Technical Manuals* were similar in scope to the *Symposium Series:* they contained groups of papers on a specific topic. But unlike the *Symposium Series* they were edited by *CEP*'s Editorial staff and contained transcripts of the discussions which accompanied the papers. Moreover, the *Technical Manuals* were largely issued in several series, for example "Safety in Air and Ammonia Plants," and "Loss Prevention," with volumes in the series appearing annually.

In shifting from the *Transactions* to *CEP* the Institute began to move the focus of its publications program from academic and research engineers to engineers working in the chemical industry. This shift had the potential of expanding readership of the flagship publication, but posed the threat of alienating academics from the society, since scholarly publications were essential to their careers. The shift also left the publications of other societies in a stronger position to capture basic contributions to chemical engineering literature. A survey in the early 1950s indicated that the American Chemical Society's *I&EC* was already publishing approximately three times more chemical engineering papers than the AIChE.[91]

To make the society more attractive to academic and research engineers and to strengthen its position in chemical engineering literature, the Institute began to consider in the early 1950s the creation of a dual publications program. Under this plan, *Chemical Engineering Progress* was to continue, and even accelerate, its shift away from the publication of heavily technical papers and continue trying to attract a broader readership by publishing more material describing and discussing current trends in industry and the profession and material of general professional interest. A new publication, in a sense a revival of the old *Transactions*, was to publish highly theoretical and abstract papers and contributions to fundamental knowledge. The new service publication, the *AIChE Journal,* was initiated in 1955 with Harding Bliss as editor.[92]

The steady expansion of the AIChE publications program can perhaps best be illustrated quantitatively. Between 1908 and 1935 the Institute published the *Transactions* annually. Each volume contained some 300 to 500 pages and averaged around 20 papers. Between 1935 and 1947 the *Transactions* were issued more frequently and grew in size, averaging around 35 to 40 papers and 700 to 900 pages. By the late 1950s, however, through *CEP*, the *Symposium Series,* the *Monograph Series,* and the *Journal,* the Institute was publishing annually an average of over 2500 pages, excluding advertisements. Of this number, approximately 1000 were *CEP* pages, which contained around three times more words to a page than the old *Transactions* pages.[93]

Meetings

In the 1940s the AIChE also expanded its meetings program. For some 35 years after its founding it had followed a policy of having two meetings a year—one near the end of the year, the other in late spring or early summer. The number of yearly meetings was increased in the 1940s, as the profession expanded, to first three and then to four. The original intention of the expansion was to have one "National" meeting and two or three regional meetings scattered over the country. But the draw of the so-called regional meetings was sufficiently high so that the term "regional" was soon dropped.[94] The original "National" meeting was soon designated as the "Annual" meeting, and the "regional" meetings were soon designated "National" meetings.

One factor which encouraged the growth of the AIChE meetings program was a major policy change. Prior to 1944 the papers presented at meetings were arranged for by the Publications Committee (until 1932) or the Papers Committee (1932-44). The latter group, in particular, applied stringent criteria which excluded all but first-class technical papers. In 1944 responsibility for soliciting papers and arranging programs for meetings was transferred to a newly-created Program Committee which was specifically charged with obtaining papers on a wider variety of subjects without regard to prospects for eventual publication.[95]

The Program Committee quickly became one of the Institute's most important groups. In 1954 it issued the first formal Program Committee manual, outlining, for the first time, the duties of all those involved in programming activities. In the 1950s it subdivided into numerous areas or divisions of interest, each headed by a chairman and responsible for guiding programming in a particular subject area. Under the Program Committee's direction both the number and variety of papers presented at AIChE meetings increased. Prior to 1940 no more than 36 papers had ever been presented in any single year and typically the figure was around 25 to 30. By 1950, however, almost 150 papers were being presented annually at meetings, and by the late 1950s more than 300. The number of presentations in previously neglected areas like marketing, process design, production control, and socio-economic matters had increased most rapidly.[96]

Throughout its first quarter century, the organization's technical contributions to chemical engineering were largely restricted to the publication of literature and to the diffusion of information through its meetings. Publications and meetings were to remain the core of the Institute's technical program through its second quarter century, but in this period it also attempted to expand its technical program through greater participation in standardization activities and through sponsorship of research.

Standards and Research

The AIChE's standards activities began much later than those of a number of other major American engineering organizations. In part this was because it was organized at a relatively late date (1908) and because its energies were at first largely absorbed by education problems. In part it was also because the American chemical industry was not very interested in standardization in the early twentieth century. By the time the AIChE was strong enough to consider seriously entering the field many of the areas of concern had already been attacked by other groups like the American Society for Testing Materials (ASTM, founded 1902), the American Society of Mechanical Engineers (founded 1880), and the American Standards Association (ASA, founded 1918). Thus the AIChE's early activities in standards remained on a relatively small scale and consisted largely of providing input to ASTM, ASME, or ASA sponsored standards.[97]

The Institute attempted to expand these activities in the 1930s, appointing in 1932 a committee to develop a list of chemical engineering symbols. This committee published a tentative list in 1938 and a revised list in 1944. Recognizing this work, the ASA Section Committee Y10 in July, 1943, appointed the members of AIChE's Committee on Symbols and Nomenclature to its Subcommittee 12 for Chemical Engineering. The result of this committee's work was the first set of standard symbols for chemical engineering, a set approved for use by the ASA in September, 1946.[98]

After this beginning the AIChE continued to participate in standards and related activities in three ways: (1) through its own Standards Committee, formed in 1954; (2) through participation in the work of the ASA, and (3) through the development of equipment testing procedures.

Of these three, the least important was the direct work of the AIChE's Standards Committee. Because other groups had achieved dominance in standards, its function was limited to reviewing the need for standards in various areas, recommending action by other organizations, and appointing representatives to committees formed by those associations.

The Institute's impact on the issuance of standards was much greater through its activities as a member of ASA. AIChE joined this body, after long debate, in 1950. By 1963 the Institute was represented on the ASA's Standards Council, on three Standards Boards (the Chemical Industry Advisory Board, the Miscellaneous Standards Board, and the Nuclear Standards Board) and on 19 sectional committees working on standards in specific areas. In addition, beginning in 1956, the AIChE assumed sponsorship of ASA's Sectional Committee N5 on Nuclear Fuel Cycle Engineering, a committee which eventually produced one of the first American standards for nuclear engineering.[99]

Although not technically a producer of standards, the AIChE's program of equipment testing procedures was clearly a "standards-like" activity. In 1944 the Institute appointed a Committee on Testing Techniques and Equipment Performance Standards, changed in 1951 to the Equipment Testing Procedures Committee. This committee's primary purpose was the development of procedures for testing chemical engineering equipment under actual operating conditions with practical, in-plant measurement techniques. The procedures were designed to secure general agreement between the sellers and users of chemical equipment on how to make performance measurements, including such areas as nomenclature, instrumentation needed for test measurements, type of data to be measured, methods of measuring data, and interpretation of results.

The AIChE issued the first of its equipment testing procedures in 1952 on heat exchangers. By the organization's "Golden Jubilee" (1958) it had issued, in addition, procedures on absorbers and condensers and was nearing completion of procedures for centrifugal pumps and mixers. The AIChE rightly considered its work in this area to be a "professional contribution of a high order," for it was an activity of value to the chemical industry and to chemical engineering educators which was not being performed by other organizations.[100]

In the early 1950s the Institute also attempted to strengthen its technical program through sponsorship of research. In 1950 Walter Lobo (Director, 1955–57), then chairman of the near-inactive Research Committee, proposed revitalizing that committee by using it as the Institute's medium for sponsoring research. The idea of a professional society sponsoring research was not new. The ASME had long had such a program.[101] But AIChE's involvement in the area was novel. Lobo proposed that the Research Committee be permitted to review engineering problems of concern to the chemical industry, solicit funds from industry to finance work on these problems, and supervise that work after it had been let out on contract. The AIChE would, ultimately, publish the results for the general benefit of the profession. Council approved the plan, and the first industry-financed, AIChE-sponsored project was initiated in July, 1952, on plate efficiencies in fractionating columns (bubble tray efficiency). Work on this project was carried on from 1952 to 1957 at three universities. The last of the three final reports was published in 1960.[102]

OTHER PROGRAMS

Most of the major changes and improvements in the AIChE in its second quarter century occurred in the two areas already reviewed in this chapter: internal organization and technical programs (publications, meetings, standards, research). The society's programs in other areas, like education, professional legislation, and employment conditions, either required less modification or were more limited.

In education, for example, the AIChE continued (in conjunction with ECPD after 1932) to accredit chemical engineering programs. Accreditation procedures were improved, and the number of accredited programs steadily rose. But there were no major new initiatives in this area.

In the field of professional legislation the organization's activities were limited and conservative, confined largely to statements and resolutions opposing the compulsory licensing of engineers. Perhaps because a larger proportion of chemical engineers work in industry and a smaller proportion directly serve the public than almost any other engineering field,[103] the Institute traditionally opposed compulsory licensing. As early as 1919, when the drive for licensing was barely underway, the Institute passed a resolution "deprecating" compulsory licensing for engineers.[104] Similar resolutions were passed every three or four years during the 1920s and 1930s.[105] In 1937 Council, while still expressing disapproval of the entire concept of licensing, voted to approve a model registration law "in principle." It agreed to endorse the 1937 model law because it required licensing only for engineers in direct contact with the public (mainly civil engineers) and because it felt that endorsement would put it in a stronger position to resist later modifications to the model law which might be "unfair" to chemical engineers.[106]

Because most registration laws contained an industrial exemption, a minority of chemical engineers was licensed by the 1950s—around 30%.[107] Since the majority of chemical engineers would have been damaged professionally by extending registration laws to cover engineers working in industry, the AIChE reacted strongly against such attempts in the 1950s, issuing several sternly worded statements. In 1959, for instance, Council declared that it was "not in favor of compulsory registration of all engineers" and that registration "of itself" did not contribute "significantly to professional responsibility and professional development."[108]

The society's activities in other political arenas during its second quarter century were likewise conservative in nature. For example, its Committee on Technological Unemployment and its Committee on the Study of Economic Conditions, both briefly active early in the Great Depression, attempted to refute the idea that technology was the cause of unemployment. They argued that Congress should study not only what technical advances deprived men of work, but also those which provided employment. In addition, as one observer noted, they came out strongly against "certain isms then current" and advised AIChE to refrain from involving itself with the National Industrial Recovery Act.[109]

The Institute's Patent Committee and Alcohol Committee continued to operate in the 1930s, regularly reporting on pending legislation in their fields, occasionally filing a protest against certain bills. But there is little evidence to suggest that these actions attracted widespread support within the organization or had any real impact on Congressional legislation.

Attempts through the Technical Manpower Committee to protect chemical engineers and chemical engineering students from the draft probably attracted more support and energy from the Institute's membership than any other political activity. But these attempts, too, enjoyed little success. In 1944 the committee reported "almost total failure in all its efforts."[110] Only in 1945, as World War II drew to a close, did it achieve a portion of its goals.

The Institute late in its second quarter century made two brief incursions into the area of economic benefits. In 1952 it conducted a salary survey as part of a general survey of membership opinion on various aspects of its organization, programs, and policy.[111] Then, in 1954, concerned with the spread of unionization and the recent formation of the Engineers and Scientists of America as a bargaining unit for engineers and technicians, Council appointed a committee to study employment conditions "conducive to improving the attitude of engineers toward their profession."[112]

This committee, chaired first by Arthur Doolittle (Vice President, 1955) and then by R. P. Dinsmore (Director, 1954–56), produced a statement on "Professional Standards" approved by the Institute in 1955. This document was rather conservative in tone, reiterating Council's opposition to unionization and declaring that the engineer's status as a professional was "endangered by reliance upon group efforts to act on his behalf in reaching short-range economic objectives." But it also contained a section outlining methods which "progressive employers" had found effective in promoting good relationships with professional personnel. To publicize its views, the AIChE published the statement in *CEP* in 1955 and reprinted it as a pamphlet, distributing it widely.[113] The pamphlet was used as a basis for discussion in a series of executive luncheons sponsored by the Institute, usually in conjunction with its meetings, in the late 1950s and early 1960s.[114] With the general prosperity existing then, however, the attraction of unions to engineers declined and interest in AIChE's pamphlet with it.

The AIChE's accomplishments in the areas of governmental interaction, professional legislation, employment conditions, and even education in its second quarter century were clearly limited. Two factors account for this. First, the Institute's energies in this period were largely absorbed by the internal organizational changes necessary to broaden its membership base and appeal. Second, the chemical engineering profession did not face any crises in the period from c1930 to c1950 comparable in severity to the legitimacy crisis of its first quarter century. Hence there were no outside forces pushing the organization to aggressively cultivate new areas. This last factor requires some explanation, for the Great Depression of the 1930s was a major crisis to some American engineering societies, pushing them to undertake significant activities in areas like professional legislation and economic conditions.

During the depression most American engineering societies faced three major problems—declining membership, declining income, and demands for increased services. The American Society of Mechanical Engineers had around 20,000 members at the onset of the depression. Its membership steadily declined in the 1930s and did not reach the 20,000 member level again until 1945.[115] The Institute of Radio Engineers had acquired 6000 members by 1931. By 1935 it had dropped to less than 5000 members.[116] The American Society of Civil Engineers had 14,218 members in 1930. In 1935 its membership was still under 15,000, and it had reached only 16,166 in 1940.[117] As membership dropped, so did income. These difficulties were compounded by member demands for new programs to alleviate their unemployment problems.

Some societies, the ASME, for instance, were so hard hit by membership and income losses early in the depression that they were unable to undertake new programs.[118] Other societies were pushed by unemployed or underemployed members into new areas. The American Society of Civil Engineers, for example, began to actively support compulsory licensing in the 1930s as a means of alleviating unemployment among its members. The ASCE also began publishing salary guidelines for employers of civil engineers and increased its political activities, enjoying some success in promoting public works legislation.[119]

But while the depression was a major crisis to some American engineering societies, either crippling them or pushing them into new professional activities, it had little impact on the AIChE. As already noted, the chemical industry and, hence, chemical engineers were not as deeply affected by the depression as other industries and other professions. Graduates in chemical engineering throughout most of the depression found work "fairly readily," according to one observer.[120] Thus, AIChE experienced no groundswell of demand for activities in areas like licensing, political interaction, and salary guidelines. It was able to concentrate on its expansion and reorganization program. The increased membership and income which resulted more than made up for the financial losses incurred by the society early in the depression. As a result, it grew steadily during the depression years, increasing from 872 members in 1930 to 1157 in 1935 and 2255 in 1940. In no year did its membership decline.

STRENGTHENING NATIONAL HEADQUARTERS

The growing size of the AIChE, the growth of its meetings and publications, and the increasing number of activities in which it was involved had begun, by 1950, to put strains on the traditional relationship between Council and headquarters staff. Traditionally, Council served as both policy maker and, along with the committees it appointed, policy implementor. It also often participated in the day-to-day management of the society. Headquarters staff from the Executive Secretary on down were expected to be relatively passive. As 1953 AIChE President W. T. Nichols observed: ". . . any attempt on the part of the paid staff to furnish leadership to a significant extent would have been viewed with alarm by many members."[121]

This arrangement had worked well when the Institute was composed of a small group of intimately acquainted executives and consultants and had a limited program and budget. It was not appropriate for a mass organization with thousands of members and a budget in the hundreds of thousands of dollars. The voluntary, part-time efforts of Council and the committees under its supervision were insufficient to handle effectively the more numerous and complex problems of the larger organization. Headquarters was unable to alleviate the situation because it was inadequately staffed and given insufficient authority.

In 1953 a Committee on the Future of the Institute, headed by G. H. Holbrook (President, 1958), pointed out that a "basic change" was needed in the organization of the headquarters staff and in the relationship between that staff and Council. Holbrook's committee recommended the addition of more full-time professionals to the staff to take the burden of program implementation from Council and its committees, to eliminate or reduce the volume of day-to-day trivia in which Council was involved, and to make membership services generally more effective. It recommended, in addition, the employment of a general manager with the broadest authority and responsibility for directing headquarters staff and implementing Council directives. Under this new arrangement, Council would serve a function analogous to the Board of Directors of an industrial corporation. It would be responsible primarily for policy making and management auditing. The Institute's Executive Secretary would be granted increased authority, responsibility, and staff for managing the organization on a day-to-day basis and implementing Council policy, making his position roughly analogous to that of the chief operating officer of an industrial organization.[122] These recommendations were supported by the 1954 Committee on the Future of the Institute, chaired by J. H. Rushton (President, 1957).[123]

The transition from a headquarters staff and an Executive Secretary with smaller responsibilities to a staff and Executive Secretary with a greater role in creating and implementing programs occurred gradually in the late 1950s and early 1960s and was to permit the Institute to consolidate the gains made in the preceding decades and even expand on them. An important step toward fulfilling the recommendations of Holbrook and Rushton was the appointment of Franklin J. Van Antwerpen to the post of Executive Secretary to replace the retiring Stephen Tyler in 1955.

Stephen Tyler had been a traditional engineering society Executive Secretary and, within those boundaries, a good one. Under his care AIChE had prospered, growing from less than 1500 members in 1937, when he assumed office, to more than 14,000 in 1955, when he left, and from 10 local sections to 48 (with 17 more in the process of formation).[124] But his authority and responsibilities had been limited by Council and by tradition.

Franklin J. Van Antwerpen, his successor, was from a different mold. He was quite familiar with the way the Institute operated, having directed its publications program since 1946. Moreover, he served under Councils which increasingly recognized that full-time professionals could provide more continuity and handle the day-to-day affairs of a large and increasingly more complex professional society better than part-time officers. Thus, during his long tenure as Executive Secretary (1955–78) Van Antwerpen was able to transform the functions of that office in accordance with the recommendations of the Holbrook and Rushton committees. Council gradually withdrew from the every day operation of the Institute, leaving it to the Executive Secretary and his staff to determine how its directives were to be implemented and to arrange matters appropriately. Furthermore, during Van Antwerpen's tenure Council came to expect its Executive Secretary to be more than a manager and to provide it with advice, guidance, and leadership. This new division of responsibilities was to work quite effectively.

COMING OF AGE

In summation, the AIChE made a large number of modifications and additions to its programs, policies, and organizational structure to broaden its membership and appeal and to increase its efficiency in the period from around 1930 to the late 1950s. It established a permanent headquarters, initiated a number of major membership drives, made significant modifications to its constitution, and enlarged its student chapter program. In addition, it established local sections all over the nation, established the principle of specialty divisions, developed a much improved and more extensive publications program, carried out significant activities in standards, and began a program to sponsor research. Then, in the 1950s, to preserve and consolidate these gains, Council modified its traditional relationship to headquarters staff, increasing the authority and responsibilities of the Executive Secretary so that AIChE's organization roughly paralleled that of a modern industrial corporation. These actions, when coupled to the general growth of the chemical industry during the same period, resulted in a steady rise in membership.

Several sets of data indicate the success of the Institute in broadening its appeal and increasing its numbers. In 1930 the AIChE was still an elite body. Nearly 90% of its members were Active members, more than 30 years old, with at least five years in responsible charge of chemical engineering operations. According to one estimate, only around 10% of all chemical engineers belonged to the Institute. By the time of its golden anniversary in 1958 the AIChE was open to all—executives, consultants, academics, mid-career engineers, and young, non-supervisory engineers. The proportion of Active members (by 1958 simply called Members) had decreased to 38%, and somewhere between 35 and 50% of all American chemical engineers belonged to AIChE, a better proportion than many of the older and larger societies.[125] In 1930 the Institute had 872 members; in 1958 it had almost 18,000.

By the mid 1950s the Institute's growth in membership and its expanding organization and technical programs were sufficient to establish it as both a major engineering society and *the* professional organization for chemical engineers. The admission of AIChE to the United Engineering Trustees (UET) in 1958 as a "Founder Society" was symbolic of AIChE's growing stature within the engineering community and of its coming of age as a professional engineering society. Although technically the term "Founder Society" means no more than part owner of UET, its real estate agent, symbolically the term has for decades meant much more. Those societies collectively called the "Founder Societies" have long been regarded as the foundation stones of the engineering profession in the United States, the organizations with the strongest claims, collectively, of representing the entire profession.

The origin of the term "Founder" dates back to 1903 when Andrew Carnegie offered one million dollars for constructing a joint headquarters building to the four main engineering societies. These four societies—later to be termed the "Founder Societies"— were the American Society of Civil Engineers, the American Society of Mechanical Engineers, the American Institute of Mining and Metallurgical Engineers, and the American Institute of Electrical Engineers. The ASCE initially refused to embrace the project, but the other three societies in 1904 incorporated the United Engineering Society (the name was changed later to United Engineering Trustees) to convert Carnegie's gift into a joint headquarters building (the Engineering Societies Building, formerly on 39th street in New York) and to provide a permanent body to administer that building and a joint Engineering Societies Library. ASCE belatedly joined the group in 1916.[126]

The American Institute of Chemical Engineers was involved in none of these transactions and certainly was not numbered among the Founder Societies for many years. In 1903–04 when Carnegie offered his gift, chemical engineering was not recognized as a distinct engineering speciality, and the AIChE did not exist. For many decades later its membership was so small and its programs so limited that it could not realistically be compared to the founder Societies in any area save education.

By the 1950s, however, that had changed, and the simultaneous need of the four Founder Societies and the AIChE for expanded office space provided the opportunity for the admission of the Institute into the United Engineering Trustees.

The AIChE had, since authorizing a headquarters in 1930, never owned office space. It had either made use of LeMaistre's office in Philadelphia (1930-37) or rented space in various structures in New York (1937 on). In the early 1950s the society began to consider raising funds to purchase a structure to house its executive offices and, in 1953, hired an independent firm to determine whether industry and AIChE members would support a drive to raise around $300,000 for that purpose. The firm's report was favorable to such a drive, but also recommended that the Institute seriously consider joining with the Founder Societies to build a new national engineering center, since the founders had outgrown their Engineering Societies Building.

This suggestion was received favorably by Council. In 1953 the Institute suspended its drive and joined with the Founders to begin studies on the location of a new national engineering center. Several joint committees made up of members from the ASCE, AIME, AIEE, ASME, and from the AIChE studied possible sites for a new structure in the mid 1950s. After very seriously considering locations in several different cities, the societies eventually agreed to erect the new building in New York at a site near the United Nations Building.

Simultaneously, the five societies conducted negotiations to determine on what terms AIChE would be admitted to the United Engineering Trustees as a fifth "Founder Society." These negotiations culminated in 1956 when AIChE offered to buy into the UET, but at a lower rate than paid by the original four. In the spring of 1958 the last of the Founders approved the document admitting AIChE to the United Engineering Trustees, and on May 1, 1958, the American Institute of Chemical Engineers officially became one of the symbolically important Founder Societies.[127]

A little more than a month later, in June of 1958, the AIChE met in Philadelphia to celebrate its golden anniversary. In formal ceremonies the American Institute of Chemical Engineers was honored for its role in establishing chemical engineering as a distinct discipline by delegates from a host of American engineering societies and from 20 foreign countries and 27 foreign professional institutions. In an address to the gathering Crawford H. Greenewalt, President of Du Pont, declared: "As a profession, we need bow to no other scientific discipline; scientifically, we have come of age."[128] The phrase "come of age" could just as appropriately have been applied to the Institute itself. In the space of three decades it had transformed itself from a small, elite engineering society into a broadly-based, professional engineering society of the first rank.

In attaining this maturity, however, it would increasingly have to consider the problem of balancing the more divergent and often contradictory interests of its now broad and diverse membership.

Chapter 2: Notes

1. Alfred Holmes White, "Chemical Engineering Education," in Sidney D. Kirkpatrick, ed., *Twenty-Five Years of Chemical Engineering Progress,* New York, 1933, p. 359; also in "The Silver Anniversary of Chemical Engineering in America" (souvenir booklet, AIChE, December, 1932) p. 46.

2. *AIChE Bulletin* (Hereafter *Bulletin*), no. 20 (1919) pp. 7-21, for the discussion on membership criteria; p. 13 for the Little quote.
3. *Ibid.*, no. 21 (1920) pp. 31-32.
4. *Ibid.*, p. 25.
5. *Ibid.*, no. 20 (1919) pp. 13-14, 18-19; no. 21 (1920) p. 30.
6. *Ibid.*, no. 21 (1920) p. 26.
7. *Ibid.*, p. 29.
8. For the general history of the American chemical industry in this period see L. F. Haber, *The Chemical Industry, 1900-1930*, Oxford, 1971, and William Haynes, *The American Chemical Industry*, vols. 2-5, Toronto, 1945-54.
9. Haber, *Chemical Industry*, p. 320.
10. Haynes, *American Chemical Industry*, v. 5, p. 43.
11. *Ibid.*, p. 31.
12. *Ibid.*
13. *Ibid.*, p. 402n.
14. United States, Bureau of the Census, *Historical Statistics of the United States, Colonial Times to 1970*, Washington, 1975, p. 674.
15. *Ibid.*
16. Albert B. Newman, "Development of Chemical Engineering Education in the United States," *AIChE Transactions*, (Hereafter *Transactions*), v. 34 (1938), supplement, p. 5.
17. *Ibid.*, pp. 5-6.
18. D. B. Keyes, "History and Philosophy of Chemical Engineering Education," *CEP*, v. 49 (December, 1953) p. 640.
19. U.S., Bureau of Census, *Historical Statistics*, p. 140.
20. *Bulletin*, no. 41 (1930) p. 6, and no. 42 (1930) pp. 25-26 (on ASME plans as stimulus for membership qualification changes).
21. For the debate see *ibid.*, no. 42 (1930) pp. 5-10.
22. Council, *Minutes*, October 17, 1930.
23. *Ibid.*
24. Council, *Minutes*, November 7, 1930. At the AIChE meeting in New Orleans in December, Langmuir announced the final results of his straw poll: 435 favored making proficiency in chemistry a requirement of Active membership, 35 opposed (see *Bulletin*, no. 42 [1930] p. 21).
25. Alfred Holmes White, "The Evolution of the Institute: The Adolescent Period," unpublished manuscript, AIChE archives, p. 15. White was President of the AIChE at the time and presided over the meeting.
26. For the debate see *Bulletin*, no. 42 (1930) pp. 9-31. The Weiss and Field quotes are from p. 19.
27. Council, *Minutes*, December 7 and December 10, 1930. The quote is from the December 10 meeting. Council's position is also mentioned in *Bulletin*, no. 42 (1930) p. 10.
28. *Bulletin*, no. 42 (1930) p. 31; Council, *Minutes*, March 10, 1931.
29. White, "Adolescent Period," p. 16, notes that it was the younger members who pushed for chemical engineering in place of chemistry as a membership requirement.
30. Council, *Minutes*, December 6, 1927.
31. Council, *Minutes*, April 12, 1929.
32. *Bulletin*, no. 40 (1929) p. 44.
33. Council, *Minutes*, September 2, 1929, and *Bulletin*, no. 40 (1929) p. 36, for the results of the questionnaire. Council, *Minutes*, June 18, 1929, for the recommendations of the ad hoc committee, and *Bulletin*, no. 40 (1929) pp. 35-47, for the committee's report and the discussion of the report.
34. Council, *Minutes*, December 1, 1929.
35. Council, *Minutes*, January 17, 1930; Van Antwerpen and Fourdrinier, *High Lights*, (Hereafter *High Lights*) p. 131.

36. Council, *Minutes*, January 8, 1937; *High Lights*, p. 131.
37. *Transactions*, v. 3 (1910) p. 44.
38. *Ibid.*, v. 4 (1911) p. 4.
39. *Bulletin*, no. 44 (1931) p. 17.
40. *Ibid.*, p. 20; Council, *Minutes*, November 1, 1931.
41. Council, *Minutes*, January 15, 1932.
42. *Bulletin*, no. 45 (1932) pp. 12, 52, 96; Council, *Minutes*, September 16, 1932.
43. Council, *Minutes*, January 15, 1932.
44. Council, *Minutes*, October 14, 1932.
45. *Bulletin*, no. 57 (1938) p. B116; no. 58 (1939) p. B40; and no. 60 (1940) p. B56.
46. Council, *Minutes*, February 11, 1938.
47. *High Lights*, pp. 37–40.
48. *Bulletin*, no. 70 (1944) pp. B65–66.
49. *High Lights*, pp. 127–28.
50. *Ibid.*, pp. 37–38.
51. *Bulletin*, no. 8 (1913) pp. 27, 48; see also no. 9 (1914) p. 10.
52. *Bulletin*, no. 15 (1917) p. 6.
53. Calvin Rice, "Fifty Years of the A.S.M.E.," *Mechanical Engineering*, v. 52 (1930) p. 269; Wisely, *American Civil Engineer*, p. 173.
54. White, "Adolescent Period," p. 6.
55. *Bulletin*, no. 23 (1921) p. 50; no. 24 (1921) pp. 47–50; Council, *Minutes*, September 14 and October 19, 1921, and September 14, 1922.
56. *Bulletin*, no. 26 (1922) p. 8.
57. *Ibid.*, no. 40 (1929) p. 24, and Council, *Minutes*, August 24, 1928.
58. Council, *Minutes*, August 24, 1928, and *Bulletin*, no. 40 (1929) pp. 24–30.
59. *High Lights*, pp. 68–70.
60. *Ibid.*, p. 68.
61. *Bulletin*, no. 57 (1938) p. B109.
62. Council, *Minutes*, December 5, 1959.
63. Wisely, *American Civil Engineer*, pp. 47–50.
64. Sinclair, *ASME*, pp. 79–84.
65. Joe B. Alford, "History of the Institute: One Hundred Years of AIME," in *Centennial Volume: American Institute of Mining, Metallurgical, and Petroleum Engineers, 1871–1970*, New York, 1971, p. 242; Charles F. Scott, "The Institute's First Half Century," *Electrical Engineering*, v. 53 (1934) p. 672; and "The Genesis of the IRE," Institute of Radio Engineers, *Proceedings*, v. 40 (1952) p. 520.
66. *Transactions*, v. 4 (1911) p. 8.
67. *Bulletin*, no. 24 (1921) pp. 76–78, 85.
68. Council, *Minutes*, January 19, 1922.
69. Council, *Minutes*, June, 1922 ("Report to Council"); *High Lights*, p. 104.
70. *Bulletin*, no. 31 (1925) pp. 10, 62; *CEP*, v. 72 (January, 1976) pp. 86–87.
71. Council, *Minutes*, December 6, 1932.
72. Council, *Minutes*, April 14, 1933.
73. Council, *Minutes*, April 13, 1934; *Bulletin*, no. 46 (1933) p. 23.
74. *CEP*, v. 57 (October, 1961) p. 97; Council, *Minutes*, May 15, 1964, p. 272.
75. *High Lights*, pp. 106–07.
76. Council, *Minutes*, November 9 and December 16, 1945.
77. Wisely, *American Civil Engineer*, pp. 47, 197, 218–21.
78. Sinclair, *ASME*, pp. 121–31, 138.
79. *High Lights*, pp. 108–10.
80. *Ibid.*, pp. 110–11.
81. Council, *Minutes*, June 18, 1927, and September 2, 1929.

82. *Bulletin*, no. 40 (1929) pp. 35-47.
83. *Ibid.*, no. 45 (1932) p. 64.
84. *High Lights*, p. 113.
85. Council, *Minutes*, March 13, 1936, and May 8, 1936.
86. On steps taken by the Institute to improve publications in the late 1930s see *High Lights*, pp. 44-45, and Council, *Minutes*, May 8, 1936.
87. Council, *Minutes*, January 11, 1946.
88. *High Lights*, p. 47.
89. *CEP*, v. 47 (October, 1961) p. 97.
90. *High Lights*, pp. 48-49.
91. "Report: Committee on the Future of the Institute," George Holbrook, Chairman, November 9, 1953 (Hereafter called the "Holbrook Report") p. 13 (AIChE Archives).
92. *High Lights*, pp. 49-50.
93. *CEP*, v. 57 (October, 1961) p. 97.
94. *High Lights*, p. 112; *CEP*, v. 49 (August, 1953) p. 74.
95. *High Lights*, pp. 46, 113-15, for the operation of the Program Committee.
96. *High Lights*, p. 46; *CEP*, v. 57 (October, 1961) p. 97.
97. On the late interest of the chemical industry in standards see *CEP*, v. 54 (December, 1958) p. 37, and Council, *Minutes*, February 18, 1955, p. 37; for early AIChE standards work in conjunction with other societies see, for instance, *Bulletin*, no. 29 (1924) pp. 20-21, 34-55, and no. 32 (1925) pp. 40-42.
98. "Letter Symbols for Chemical Engineering," pamphlet, New York, American Institute of Chemical Engineers, June, 1956. This pamphlet has a brief review of AIChE involvement with chemical engineering symbol standards. See also, *High Lights*, pp. 155-56.
99. The best review of AIChE standards activities in the 1950s and 1960s is the report of the ad hoc Committee on Standards Activities in Council, *Minutes*, September 28, 1963, Appendix A, pp. 347-88. See also, *High Lights*, pp. 155-56.
100. *High Lights*, pp. 144-46, and *CEP*, v. 77 (March, 1981) pp. 44-48, for reviews of the equipment testing procedure program. The quote comes from Council, *Minutes*, September 28, 1963, Appendix A, p. 388.
101. Sinclair, *ASME*, pp. 158-59.
102. *High Lights*, pp. 121-25.
103. Engineers' Joint Council, *The Engineering Profession in Transition*, New York, 1947, p. 13. The EJC survey indicated that 91.9% of all chemical engineers were privately (largely industrially) employed, a higher figure than any of the other major engineering fields.
104. *Bulletin*, no. 20 (1919) pp. 44-47, 62-64.
105. For instance, *Bulletin*, no. 26 (1922) pp. 72-73; Council, *Minutes*, February 9, 1927; June 12, 1931; and October 14, 1932.
106. Council, *Minutes*, September 10, 1937.
107. *CEP*, v. 49 (October, 1953) p. 56.
108. *CEP*, v. 55 (July, 1959) p. 17; Council, *Minutes*, May 16, 1959, pp. 271, 290-91.
109. *High Lights*, p. 142; see also *Bulletin*, no. 44 (1931) pp. 48-49.
110. *Bulletin*, no. 70 (1944) p. B113.
111. "National Survey Questionnaire: A Report to the Members," *CEP*, v. 49 (April, 1953) pp. 17, 28, 56; (July, 1953) pp. 17-18, 20, 22.
112. Council, *Minutes*, September 24, 1954, p. 110.
113. "Professional Standards," *CEP*, v. 51 (November, 1955) pp. 50-52.
114. *High Lights*, p. 76.
115. Sinclair, *ASME*, pp. 176, 201.
116. "The Genesis of the IRE," Institute of Radio Engineers, *Proceedings*, v. 40 (1952) p. 520; Laurens E. Whittemore, "The Institute of Radio Engineers—Fifty Years of Service," *ibid.*, v. 50 (1962) p. 540.

117. Wisely, *American Civil Engineer*, pp. 54, 423.
118. Sinclair, *ASME*, pp. 176, 179, 194, and 157-95 passim.
119. William Wisely, "Professional Turning Points in ASCE History," *Civil Engineering—ASCE*, New York, [1978], pp. 164-66.
120. Newman, "Development of Chemical Engineering Education," p. 5.
121. *CEP*, v. 49 (June, 1953) p. 68.
122. "Holbrook Report," pp. 1-3, 7-12.
123. Council, *Minutes*, December 11, 1954; for the report of J. H. Rushton's committee see pp. 178-191.
124. For appreciations of Tyler's work see *CEP*, v. 50 (December, 1954) p. 96, and v. 62 (May, 1966) p. 184.
125. The 10% estimate for 1930 is taken from a graph in the "Holbrook Report" of 1953, fig. 1. The figure of 39% is from *CEP*, v. 58 (January, 1962) p. 13; the estimate of 47% is from Gloria M. Lambson, "Forecasting Employment Demand for Chemical Engineers," Master of Business Administration thesis, New York University, October, 1976, p. 19.
126. Layton, *Revolt of the Engineers*, pp. 82-84, and Ellis Rubinstein, "IEEE and the Founder Societies," *IEEE Spectrum*, v. 13 (May, 1976) pp. 76-77.
127. *High Lights*, pp. 132-35, provide a good review of the events surrounding AIChE admission to the United Engineering Trustees (the Founder Societies).
128. *AIChE Annual Report*, 1958, p. 3.

CHAPTER 3

Expanding Programs, c1958–c1970

In the dozen years immediately following admission to the Founder Societies the AIChE consolidated its claim to being an engineering society of the first rank. The organizational mechanisms and technical programs that had played a major role in expanding the Institute's membership in the preceding decades were refined and expanded. At the same time, the AIChE began to increase, although initially in a rather modest manner, its activities in areas (e.g., employment conditions) which had been somewhat neglected in the Institute's second quarter century when its energies had been largely absorbed in the extension of its technical programs and working structure.

INTERNAL ORGANIZATION

The Institute's continuing concern with internal organization, as well as with its technical programs, can be explained rather easily. Even though its internal organization paralleled those of other major societies in many ways, it had not reached the same stage of stability and maturity. There were still problems to be ironed out. These included

- inadequate headquarters facilities,
- excessive electioneering,
- the unclear role of technical divisions, and
- very loose ties between local sections and the national organization.

Headquarters Facilities

Between 1937 and 1958, as noted in the previous chapter, the AIChE had leased headquarters space, moving every five to ten years to new quarters as more space was needed. The admission of the Institute to the United Engineering Trustees in 1958 as a Founder Society and the decision of the Founders to erect a new 18-story structure (the United Engineering Center) near the United Nations was to end the society's semi-nomadic existence and provide the Institute with quarters that were more roomy, more permanent, and more prestigious, as well as quarters which it partially owned.

Funds for the new structure were raised from the sale of the old Engineering Societies Building ($2 million), gifts from industry ($5 million), and contributions from individual members of the Founder Societies ($3 million). To raise the $3 million in individual contributions each of the societies was assigned a quota based on size.[1] AIChE's quota was $300,000. The Institute began soliciting funds from its membership, largely through its local sections, in the summer of 1958. Before the end of the summer of 1959 the AIChE had reached its quota, easily the first of the participating societies to do so.[2] Ground breaking for the new United Engineering Center took place on October 1, 1959. The structure was completed in 1961, providing the Institute with what has been, ever since, its home.

Elections

The growing size of the Institute not only created problems with housing its offices, it also created problems with election procedures. As long as the organization was small and most members knew one another, elections had not been a major problem. All Active members were permitted to send in nomination ballots, and candidates listed on more than ten nominating ballots were placed on the election ballot. Most of those voting knew the qualifications of the candidates. Distribution of campaign literature was not necessary. This system, however, had begun to require modification as the Institute expanded its membership base and had required additional modification as it grew larger and extended the franchise to Associate members. In 1954, for instance, the society abandoned the use of nominating ballots (a system revised several times since 1908), replacing them with a Nominating Committee. To preserve the tradition of open, contested elections, however, the new Nominating Committee was required to submit to the voters at least two candidates for the office of Vice President (since the 1920s traditionally the President-elect) and for every vacant Directorship on Council. The nomination of additional candidates by petition was made relatively easy: only 50 signatures were required (raised to 100 in 1965). Brief biographical sketches published in *Chemical Engineering Progress* (*CEP*) and enclosed with the election ballots provided voters with information about the candidates.[3]

The decision to continue multi-candidate elections kept interest in Institute election campaigns relatively high, but it also created problems. Because the elections were contested, electioneering had become, by the 1960s, a regular part of election campaigns. Candidates and their supporters often sent out mass mailings to the membership, distributed campaign posters to local sections, and made phone calls to seek votes. By the late 1960s active electioneering had become "essentially a must for all candidates" and had reached the point where it was, in the view of some AIChE members, "not in the best interests of mature, professional, business-like procedures." Walter B. Howard (Director, 1969-71) thus proposed in 1969 that it, in imitation of most other major engineering societies, eliminate all electioneering. He argued that the brief biographical sketches published in *CEP* and enclosed with the ballots contained sufficient information.[4]

Council, however, rejected this approach, maintaining that election campaigns and the interest shown by candidates in them were a sign of "professional vitality." To refine the system Council attempted to lay down more stringent election guidelines in the late 1960s. These guidelines prohibited local sections, committees, and divisions from acting as units in behalf of candidates and forbade election deals between units of the Institute. They also discouraged the use of mass mailings and urged that election campaigns be conducted in good taste, inexpensive, and carried out at a high professional level.[5]

Divisions

The AIChE had just begun to establish a divisional structure at its 50th anniversary. In 1958 it had two divisions: Nuclear Engineering (1954) and Heat Transfer (1958; now called the Heat Transfer and Energy Conversion Division). But the Institute had established no standard procedures for the creation of new divisions and no performance standards for the existing ones. Moreover, although the leadership generally recognized that the divisions should have some responsibilities for planning programs at national meetings and some autonomy in meeting these responsibilities, the relationship between divisions and the national Program Committee had not been formalized. This situation did not provide sufficient flexibility in meeting the needs of technical specialties.

To work out the respective programming responsibilities of divisions and the Program Committee, Council appointed an *ad hoc* committee headed by S. W. Churchill (President, 1966) in 1962.[6] The committee, in addition to recommending creation of a strong planning board (eventually the Executive Board of the Program Committee) to govern the now large and cumbersome Program Committee, recommended creating additional divisions with programming responsibilities and requiring all divisions to form their own program committees and hold at least one program annually at national meetings. Council approved these recommendations in late 1963 and appointed an *ad hoc* committee chaired by Ted Burtis (President, 1967) to develop an orderly mechanism for the creation of new divisions.[7]

This committee made its final report to Council in December, 1964. Although some AIChE members feared that an expanded divisional structure would weaken the national organization, the committee advocated the immediate expansion of the Institute's divisional structure through the formation of a Petroleum & Petrochemicals Division; an Education Division; a Pulp, Paper, & Wood Products Division; and a Food & Biochemical Division. The committee also suggested that Council in the near future consider creation of a Manufacturing Division, an Engineering Sciences & Fundamentals Division, an Aerospace Division, an Environmental Division, and a Communications Division. Finally, it submitted a set of standardized procedures, similar to those of other engineering societies, to guide the formation of divisions. These procedures required, first, the establishment of a technical committee in an area. After a technical committee had operated successfully for a year, presented at least one symposium at a national meeting, and secured membership pledges from 100 members, Council could elevate it to divisional status. To insure national control over divisions, Council was to approve membership requirements, dues, and annual budgets.[8]

Council approved the recommendations of the Burtis Committee, and in 1965 the Institute's Members voted favorably on modifications to the constitution required to implement the recommendations. W. B. Franklin, the Institute's 1965 president, made the development of the divisional structure recommended by the Burtis Committee one of the goals of his administration. Two of the recommended divisions emerged with relative ease: the Petroleum & Petrochemicals Division (now Fuels & Petrochemicals Division), thanks to the efforts of Hugh Guthrie (President, 1969), and the Food & Biochemical Division (now Food, Pharmaceutical, & Bioengineering Division), thanks to the efforts of Don Dahlstrom (President, 1964). Efforts to form a Pulp & Paper Division and an Education Division, however, encountered difficulties.[9] It was not until 1968 that a Forest Products Division was established. In 1968 Council also approved the creation of a Materials

Engineering & Sciences Division, and in 1970 it approved the creation of an Environmental Division. Thus between 1964 and 1970 the Institute fleshed out its divisional structure, expanding from two to seven divisions.

Several of the new divisions quickly established very active programs, for instance, the new Petroleum & Petrochemicals Division. In 1967, in its first year of operation, the Petroleum & Petrochemicals Division not only more than fulfilled its programming responsibilities, but also established a newsletter. In 1968 it strengthened its program committee and in 1969 initiated a divisional award and formed a local section activities committee. By the early 1970s it had become one of the most active of AIChE's divisions, presenting, for instance, 25 sessions at the Dallas National meeting in February, 1972. That same year the Petroleum & Petrochemicals Division created a program review committee to study means of improving the quality of divisional papers.[10]

The expansion of the Institute's divisional structure and the standardization of procedures for organizing divisions occurred with relative ease. There was a general consensus that the modifications made were needed. This was not the case with the society's other major organizational refinement of the 1958-70 period: the revision of membership requirements for local sections.

Local Sections

In the 1940s and 1950s the Institute's local sections had rapidly increased in number: from 12 in 1940 to 37 in 1950 to 77 in 1960.[11] But this growth, while encouraged by the national organization, had begun to create strains. A number of them had become almost independent clubs, with few ties or loyalties to the national organization.[12] Two things encouraged this attitude. First, formal ties between the national society and the local sections were weak, largely limited to national providing a local section officers' manual, mailing lists, publicity in *CEP*, and the Speaker's Bureau.[13] Second, unlike the other Founder Societies the AIChE did not require members of local sections to be members of the national organization. Thus many chemical engineers joined only the local section without joining the national organization. In 1957 approximately 27% of local section membership was made up of non-national members. Even more critical, non-AIChE members composed the majority in a number of sections (16 in 1961).[14]

Between 1956 and 1965 the Institute took a number of steps to remedy the problems created by the rapid and largely unsupervised expansion of its local section structure and to fashion stronger ties between local sections and the national organization. In 1956, for instance, Council extended the liaison concept, used for some years previously to insure communication with Institute committees, to local sections.[15] Under this system Institute committees, divisions, and local sections were divided among Council members. Each Director and Officer was expected to familiarize himself with the committees, divisions, and local sections assigned him, contact and serve as advisor to these groups, bring their concerns before Council as a whole, and, generally, provide a specific point of contact with the national organization.

In May, 1959, the Institute issued a local section organization manual to better define the relationship between the national office and its local sections. At the same time Council issued model bylaws for local sections which prohibited non-AIChE members of local sections from voting and holding office.[16]

Feeling that these were insufficient remedies for the problem, Council began to consider stronger measures at its December, 1960, meeting. Irv Leibson (President, 1974), chairman of the Membership Committee, reported that the question of requiring all local section members who did not belong to the Institute to join had been discussed at two Local Section Activities meetings during the year and asked Council for a discussion of the merits of the idea. He warned, however, that such a requirement was "not particularly popular" with local sections. After considerable discussion, Council voted to ask its Constitution and Bylaws Committee to prepare provisions to require that all members of local sections be members of the national organization by January 1, 1965, but to permit new local sections to admit non-members of the national organization during the first three years of their existence.[17]

This proposal encountered severe opposition from AIChE's local sections. They argued that non-national members forced out would be lost to the Institute forever, that non-national members were often crucial to the health of local sections, and that they should be allowed a period of time to solve the problem in their own way. In February, 1961, Council, following extended debate, reiterated its conviction that all local section members should be members of the Institute, but agreed to defer action to give them time to put their houses in order.[18]

Unfortunately, this did not occur. John McKetta's September, 1961, survey of local sections indicated that 31% of their members were not AIChE members, up from 27% four years earlier.[19] A second survey, conducted by W. R. Marshall in May, 1962, indicated that very few sections had decreased their proportion of non-national members and that many were not planning to modify their bylaws to conform to Council's desires. Thus Council in May, 1962, modified the Institute's bylaws to require all non-AIChE members of local sections to join the national organization after a grace period. Council asked all local sections to modify their bylaws accordingly by January 1, 1965.[20]

To ease the impact of the new requirement and to soothe the wounds created by this action, the Institute's leadership between 1963 and 1967 initiated a number of programs to improve relations with local sections. The first of these programs was a local section officers' training program. The crisis over non-national membership in local AIChE sections came at the same time that Council was considering a dues increase. This coincidence created very real fears that the combined impact of the Institute action against maverick sections and a dues increase might seriously hurt the Institute. To explain the need for the dues increase and to try to improve relations with local sections, 1962 president John McKetta visited every AIChE local section.[21] His successor, W. R. Marshall, Jr., without hope of beating this record, adopted another expedient: bringing local section officers together on a regional basis (later done on a national rather than a regional basis) for day-long presentations and workshops.[22] The first local section officers' day in March, 1963, was received enthusiastically by participants.[23] This innovation, extended to two days in 1965 and enlarged in 1970-71 to include division and committee officers, became a permanent annual event in the Institute and an essential element for integrating local sections (and later divisions) into the national organization.

To further cement relations between the national organization and the local sections Council began to consider in 1963 a dues rebate program. Dues rebates to local sections were common among other engineering societies,[24] and they had been considered by AIChE on earlier occasions, for instance in 1945.[25] The issue was revived in 1963 as part of national's efforts to persuade local sections to limit their membership to national AIChE members. The proposal faced serious opposition since national members not active in local sections and the larger, financially-prosperous sections did not favor it. A 1963 poll indicated only 38% of AIChE members favored dues rebates, while 49% opposed.[26]

In an effort to reach a compromise the Institute's *ad hoc* Committee on Local Sections, chaired by Don Dahlstrom (President, 1964), recommended in December, 1963, that the Institute provide only "service rebates." That is, national headquarters would increase its services to local sections (for example, by having a full-time secretary for local section affairs and by funding speakers' tours), but would not provide cash rebates.[27] The compromise proved unsatisfactory. Local sections had little interest in "service rebates."[28]

In 1964 the ad hoc Committee on Local Sections, now chaired by W. B. Franklin (President, 1965), proposed, instead, a cash dues rebate program with a sliding scale based on the size of local sections. This program required them, in return for the rebates, to use part of their rebate monies to mail local section program and meeting information to all AIChE members in their areas at least several times each year. Council accepted this system in December, 1964, and, at the same meeting, authorized its Executive Secretary, Van Antwerpen, to revoke the charters of all local sections not complying with the new bylaw requirement by January 2, 1965.[29]

In addition to providing local sections with rebates and with increased guidance (through the local section officers' meetings), the Institute also increased its role in local programming through an expansion of its Speaker's Bureau. This service, initially established in the late 1940s, provided sections with lists of good speakers willing to speak for expenses only. Traditionally run out of the Executive Secretary's office, it was placed in the charge of Harold Abramson, new head of the Institute's Continuing Education Department, in 1966. He put the operation of the service on a more efficient basis and extended the practice of organizing speakers' itineraries so that they could speak to several closely adjacent local sections, permitting these sections to split costs. In his first year, 1966–67, the Speaker's Bureau provided seven speakers who delivered 35 speeches to 30 sections. The level of activity increased rapidly. In 1969–70 the bureau provided 33 speakers, who delivered 118 speeches to 56 sections. The bureau continued to operate at this level or slightly higher throughout the 1970s, providing programs for at least 10–15% of all local section meetings held in any given year.[30]

McKetta's tour of all local sections; the initiation of local section officers' days; the requirement that all local section members (after a grace period) belong to the national organization; the development of a rebate program; and the creation of a stronger speaker's bureau contributed to welding the Institute's local sections to the national organization during the 1960s. In fact, the creation of stronger ties between local sections and the national organization was perhaps the Institute's most important and successful organizational refinement of the period.

Other Organizational Modifications

One of the Institute's programs which did not reach fruition was its attempt to create a mechanism for securing funds from industry on a regular, annual basis. In the early 1960s the Institute was considering several proposals to solicit funds from industry for special projects (a career guidance film, various research programs). In February, 1962, Council appointed an *ad hoc* Committee on Industrial Funds to consider whether the Institute should continue trying to raise funds from industry through special solicitations for particular projects, or attempt to raise funds through a single large fund drive or some type of sustaining membership.[31]

In May, 1962, this committee recommended that the organization establish a system to permit industry to contribute to AIChE activities on a "yearly dues basis" through an "Industrial Associates" plan. Industrial Associates would pay the Institute annual dues varying between $500 and $5000, depending on company size. In return for the dues, the AIChE would initiate a number of projects hopefully of value to industry. New publications, refresher courses for working engineers, and production of a guidance film were among those mentioned. Participants were also to receive one set of all AIChE publications for each $500 they paid in dues.[32] Council tentatively approved the project in May, 1962, and gave final approval in August, 1962.[33]

Reception of the plan by industry was less than enthusiastic. By May, 1964, only nine companies, with a total contribution of only $11,500, had become industrial associates.[34] A new promotional letter was drafted emphasizing educational programs as the primary intended use for dues gathered from industry, but it did not help. By January, 1965, the plan had raised only $18,000, too little to have a significant impact on continuing education.[35] In February, 1965, Council abandoned the Industrial Associates plan on an eight to four vote.[36]

Why the AIChE's plan to raise funds from industry on a sustained annual basis failed is not certain. Institute officials strongly felt that the problem had not been in their "sales" effort. They felt, instead, that industry was not accustomed to looking to technical societies to solve problems like continuing education, the primary *quid pro quo* being promised by the Institute in return for industrial support.[37] Industry's fear that becoming an "Industrial Associate" of AIChE would lead to a myriad of requests from other technical societies for similar associations was, however, probably of equal importance.

With the exception of the Industrial Associates plan, the AIChE's attempts to refine and strengthen its organizational structure between 1958 and 1970 were largely successful. In practically all areas—housing, elections, divisions, local sections—the Institute's internal working structure was stronger in 1970 than it had been a dozen years earlier. In no area, however, with the possible exception of relations with local sections, was the increase in strength greater than in national headquarters itself. Between 1955 and 1970 the efficiency of national headquarters increased significantly. Much of the credit for this increased effectiveness belongs to Van Antwerpen, the Institute's Executive Secretary. He was, for instance, largely responsible for the decision to purchase a data processing system in 1957.[38] This system, the first of its kind to be used by an American engineering society, made the retrieval and updating of membership information much more efficient. In addition, he assembled in the late 1950s and in the 1960s a staff of skilled professionals to take direction of many Institute activities. For instance, in 1957 Joel Henry was hired as Assistant Secretary and placed in charge of organizing AIChE meetings, a task he performed very effectively for over two decades. In 1959 Larry Resen was brought in as editor of *CEP;* in 1966 Harold Abramson took charge of the Speaker's Bureau and the newly-created Continuing Education Department. Two other long-term members of the Institute's staff were also to play a major role in improving Institute services during its third quarter century: Sylvia Fourdrinier, who served as Assistant Secretary in charge of membership services and as managing editor of the *Symposium Series* and the *AIChE Journal,* and Harold Hansen, the Institute's Controller. In comparing the AIChE's national staff to those of other engineering societies, the 1967 *ad hoc* Committee on the Future of the Institute concluded: "Our Executive Secretary has done an excellent job."[39]

Besides establishing itself in the United Engineering Center, imposing more stringent rules on electioneering, solidifying its divisional structure, improving relations with local sections, and strengthening its headquarters staff, the AIChE made a number of other organizational refinements in the late 1950s and in the 1960s. In late 1962 Van Antwerpen and Marshall (President-elect for 1963) held the first presidential retreat and planning session, an extended meeting between the Institute's Executive Secretary and the President-elect to review the problems faced by the organization and to plan for the upcoming year. Because advanced planning is almost a necessity for a president to be effective during his one-year tenure in office, these retreats quickly became an annual practice. In 1970 the Institute created a new membership grade—Fellow—to recognize significant professional contributions among chemical engineers in active practice for at least 25 years and full Members of the Institute for at least 10 years.[40]

At the same time the Institute was refining its organizational structure, it continued to expand and strengthen its two oldest technical programs: meetings and publications.

TECHNICAL PROGRAMS

Meetings

Between 1958 and 1970 the AIChE continued to hold four meetings annually, distributing them widely for geographical balance. While the rate of growth in meeting attendance did not match the rate of growth of the Institute as a whole, it was significant. In the late 1950s attendance at the AIChE's four regular meetings averaged slightly over 5000; in the early 1960s the average was slightly under 6500; and by the late 1960s the average was around 8000. The number of sessions held annually at these meetings grew even more rapidly. In the late 1950s only around 80 to 85 sessions were held annually. By 1970 there were over 200.[41] These figures do not include meetings sponsored in conjunction with other societies. There were many of these, both nationally and internationally. Nationally, for instance, the AIChE's Heat Transfer Division continued to hold a joint annual Heat Transfer Conference with the American Society of Mechanical Engineers. Internationally, almost every year from the early 1960s on the AIChE participated in or jointly sponsored meetings with foreign or international chemical engineering organizations such as the Institution of Chemical Engineers in London or the InterAmerican Confederation of Chemical Engineering.

The growth of the Institute's meetings created a number of problems. Organizationally, too much responsibility was placed on the shoulders of the chairman of the national Program Committee or on the members of the unofficial "Program, Policy, and Planning Committee" which had emerged within that committee. They were expected to plan programs, coordinate them, and take care of administrative details (correspondence with authors, arranging for meeting rooms, etc.). In September, 1963, an *ad hoc* Committee on Programming, chaired by S. W. Churchill, suggested that Council formally establish a board at the national level to take responsibility for the planning and conduct of meetings, in effect legitimizing the unofficial "Program, Policy, and Planning Committee" as the Executive Board of the Program Committee. The *ad hoc* committee also recommended that a substantial portion of the Program Committee's responsibilities for planning and guiding programs be transferred to the Institute's divisions and that responsibility for many of the details of program organization be transferred to the national headquarters staff.[42] Council approved the recommendations in January, 1964.[43]

The most important new addition to the meetings program between 1958 and 1970 was the AIChE Petrochemical and Refining Exposition. The idea of an exhibition to enable sellers of capital equipment and services to present their wares to buyers in the chemical processing industry was not new. For some decades the Exposition of Chemical Industries (Chem Show) had been held biennially in New York. But there had been no exhibition specifically tailored to the petrochemical and refining industry or held on the Gulf coast where that industry was centered. In 1957 Van Antwerpen broached the idea of Institute sponsorship of an exposition as a means of increasing income and providing a needed service to the manufacturers and buyers of chemical processing equipment. But implementation of the idea was delayed due to planning for the AIChE's 50th anniversary celebration. In October, 1959, Council voted $35,000 to establish a petrochemical exposition.[44] Shortly after a professional firm was hired to manage the show.

The first AIChE Petrochemical and Refining Exposition was held in New Orleans in conjunction with the February–March, 1961, AIChE meeting. It was "an unqualified success."[45] Combined registration for the exposition and the national meeting was 3248, a new record for an AIChE-sponsored event. Registration for the meeting alone was 2481, third highest in AIChE history.[46]

With this success, the semiannual exposition became a regular part of the Institute's meetings program, held in conjunction with a February or March national meeting. Both registration and the number of exhibitors grew steadily through the 1960s. The 1967 and 1969 expositions held, respectively, in Houston and New Orleans drew more than 225 exhibitors and more than 14,000 registrants each. These expositions, moreover, continued to have a positive impact on meeting attendance. AIChE meetings held in conjunction with them drew between two and three times more registrations than national meetings held in February or March in years when Expos were not held.[47]

Although the proportion of Institute membership attending its national meetings tended to decline in the 1960s, something almost inevitable as the Institute continued to grow, those who did attend were generally satisfied with the organization's efforts to keep its technical meetings at a high caliber. After the 1972 New York meeting, 15% of those polled rated the technical program as excellent, 64% as good. Around 75% agreed that the meeting sessions were helpful to their jobs, and 81% felt they were helpful for professional development.[48]

The Institute's publications program, like its meetings program, was improved and expanded in the period from 1958 to around 1970.

Publications

On the eve of its 50th anniversary the Institute published three primary periodicals: *Chemical Engineering Progress* (*CEP*), the flagship publication, on a monthly basis; the *AIChE Journal*, devoted to articles on new research and theory, quarterly; and the *Symposium Series,* collections of related articles on specific subjects of narrow interest, at irregular intervals. This arrangement was quite practical. Almost all publishable manuscripts fit readily into one or the other of these periodicals.

The most important of the publications was *Chemical Engineering Progress,* the official organ of the organization, the publication which was automatically provided to all members of AIChE and the source of most of the Institute's advertising revenue. *CEP* faced the very difficult task of trying to provide for both the technical and non-technical needs of AIChE members, while at the same time serving as an outlet for institutional news and information. It enjoyed success in adjusting to these circumstances. Under Larry Resen, who became editor of *CEP* in 1959, the proportion of narrow, highly technical articles published in *CEP* continued to decrease, while the number of articles appealing to a broad readership (for example, articles on career development, engineering pension problems, engineering manpower) steadily increased. As a result the periodical gained in support among Institute members. A 1963 survey, for instance, indicated that 64% of AIChE members regarded the publication quite favorably, while only 11% were actively dissatisfied. The same survey indicated that among AIChE members it was the most widely read chemical engineering journal.[49] A 1975 poll of 77 prominent AIChE members indicated that 79% were satisfied with *CEP,* with a significant number of these praising the editor's ability to alter format and emphasis to match the changing needs of membership.[50]

The use of *CEP* as a medium for the exchange of computer programs provides one example of the attempts made in the years following the Institute's golden anniversary to increase the flagship publication's timeliness and utility to the chemical engineering profession. In the late 1950s computers were just beginning to come into widespread use in industry, but, due to lack of contact and proprietary restrictions, there was considerable duplication of effort between companies attempting to make use of the new technology.

In 1958 the Institute's new Machine Computation Committee, which was to be very active in the late 1950s and early 1960s in promoting workshops on the use of computers and in private program interchange, proposed that AIChE sponsor an industry-wide interchange of computer programs to reduce this duplication. Abstracts of programs submitted by industry or individuals to the Machine Computation Committee and accepted were to be published in standardized form on a regular basis in *Chemical Engineering Progress,* with those submitting abstracts agreeing to compile full program manuals for publication if sufficient demand for a program existed.[51]

CEP published the first set of abstracts in May, 1959.[52] The growth of program exchange through users groups and vendor program libraries, and the growing availability of economical programs and program services through private companies eventually eliminated the need for this type of activity. But before this happened in the mid 1960s, *CEP* had published more than 100 abstracts, and the Institute had published over a dozen computer manuals.

Another effort to increase the utility and timeliness of *CEP,* as well as other AIChE publications, was the introduction of an information retrieval system. A workable system, initially developed at Du Pont, was made available to the Institute in 1960 through the efforts of Raymond P. Genereaux (Director, 1958–60). Adapted by a subcommittee of the Standards Committee to AIChE literature, the system had three elements: a special thesaurus, a keyword index, and an abstract. The *Chemical Engineering Thesaurus* developed by the subcommittee provided a list of standardized terms used in chemical engineering, their synonyms, and their relationships. Authors contributing to *CEP* and other Institute publications were asked to supply with their contributions key indexing words taken from the thesaurus and an abstract. The abstract and index words were then published with the contribution so that readers could clip and use them in personal filing systems.[53]

CEP began to use the information retrieval system in June, 1961, and continued to use it regularly for more than a decade. For a period it enjoyed modest reader support. A survey published in 1965 indicated that about a third of CEP's readers used the abstracts, about 10% specifically for personal retrieval systems.[54] The system was soon copied by other groups. Shortly after AIChE put it into operation, the Engineers' Joint Council adopted it, with slight modifications, and prepared a more general thesaurus of engineering terms. By 1964 some modification of the AIChE system was in use by AIME, ASCE, ASTM, IEEE, and several other major American engineering societies.[55]

Like the computer program abstracts, the AIChE information retrieval system's life was limited. It was dropped in the mid 1970s. Several factors contributed to this. By the mid 1970s the system had become somewhat obsolete. The thesaurus, for example, was badly in need of updating. Moreover, the system had been designed primarily for mechanical retrieval systems; by the mid 1970s computerized retrieval systems had emerged. Finally, because authors often did not use the thesaurus properly (if at all), the author-generated indexes were frequently of poor quality. In brief, maintenance, updating, and quality control in the system were insufficient.

The other two primary elements of the Institute's publications program—the *Symposium Series* and the *AIChE Journal*—were designed for a different market and made other contributions to the program. The *Symposium Series,* despite wide variations in subject matter, was self-supporting and contributed significantly to the Institute's ability to respond to the growth of technical specialties by serving as an outlet for the publication ambitions of AIChE's technical committees and divisions. For instance, the AIChE's Water Committee began publishing an annual volume in the series on *Water* beginning in 1967 and continued to publish the annual volume after it had become a major section of the Environmental Division. In 1970 the AIChE published the 100th volume in its *Symposium Series,* and by 1983 it had published almost 225 volumes.

The *AIChE Journal* served a different function for the society. A prestigious magazine for the publication of research, it was designed more for those interested in the fundamentals of chemical engineering (and academic chemical engineers in particular) and not for broad readership. It served this function well. A 1974 review of AIChE publications by the Publications Committee indicated that the *Journal* was superior to competitive publications for "reviews of knowledge."[56] Moreover, the *Journal* steadily acquired an international reputation. For instance, Trevor Evans, General Secretary of Britain's Institution of Chemical Engineers, has called it the world's premier chemical engineering research journal.[57]

Nonetheless, the more limited market for which the *Journal* was designed and the highly abstract nature of many of its articles created some difficulties. In addition the *Journal* also experienced periodic problems with publication backlogs.

The Institute countered these difficulties in several ways. It balanced the highly theoretical content of the *Journal* by moving the editorial content of *CEP* steadily towards broader, less technical articles which appealed to broader readership. The publication backlog was reduced by expanding the *Journal* from a quarterly to a bimonthly beginning in 1960 and by increasing its size, editing more severely, and imposing a more stringent acceptance policy in the early 1970s. These steps improved the quality of the publication and insured its continuance as an important element of the AIChE publications program.

In addition to maintaining and improving its existing publications, the Institute attempted to expand its publications activities in several directions in the late 1950s and early 1960s. The results were mixed. The short-lived publication of computer manuals has already been mentioned. Beginning in 1958, in an attempt to more closely involve members in Institute affairs, the AIChE also began publication of an annual report. It survived only five years and was abandoned in 1963 when membership showed little interest and the Institute needed funds elsewhere.[58] At about the time the Institute began publishing annual reports, it introduced the first of its *Technical Manuals*. These were similar to the volumes in the *Symposium Series*. They contained collections of articles on specific topics. They differed from the *Symposium Series*, however, in several respects. The *Technical Manuals* were edited by the *CEP* staff; they included transcripts of the discussions which accompanied papers, as well as the papers themselves; they were more oriented to practice; and most of the *Manuals* were published in several series on a regular annual basis (for example, *Safety in Ammonia Plants*).

Another new element of the Institute's publication program was *International Chemical Engineering*, a new quarterly journal initiated in October, 1961, to provide English-speaking engineers with access to chemical engineering work in Communist-bloc (and later other) countries through translations of articles. Funded with the aid of a grant from the National Science Foundation and edited initially by Waldo Hoffman, *International Chemical Engineering* did not attract widespread member support in the form of subscriptions, but it provided a valuable service, was self-sustaining, and, therefore, became a continuing element of AIChE's program.[59]

Standards and Research

In the two technical fields of meetings and publications the AIChE clearly expanded and improved its programs between 1958 and 1970. In two other technical fields—standards and research—the opposite occurred. In both of these areas the Institute cut back on its commitments in the 1960s.

Several factors explain this divergent performance. First, there was general consensus among both the Institute's leadership and members that meetings and publications were vital programs. There was no widespread agreement about the need for AIChE sponsorship of standards and research. Second, the Institute's publications and meetings programs were sustained by a full-time professional staff; the standards and research programs were much more heavily dependent on time donated by part-time volunteers. Finally, there were no bodies outside of AIChE to promote as effectively as the Institute the dissemination of chemical engineering knowledge in all of its aspects—theoretical and practical—through a combined program of meetings and publications. There were bodies outside of the Institute better equipped and much more experienced in carrying out standardization work (the American Standards Association, now the American National Standards Institute) and research (universities and industry). It is, therefore, not surprising that the Institute's nascent programs in standards and research did not flourish to the same extent in the 1960s as its more mature and vital programs in meetings and publications.

EXPANDING PROGRAMS 65

By the early 1960s the AIChE's leadership had begun to recognize that groups devoted exclusively to standards, like the ASA would soon dominate the standards field.[60] The only engineering societies able successfully to carry out the independent development of codes and standards in the 1950s and 1960s were those, like the ASME, which had initiated such activities prior to the founding of ASA in 1918 and which utilized full-time staff instead of volunteers in the area. And even those groups were frequently choosing to work through ASA instead of attempting to develop standards of their own.

Faced with these considerations and the high cost of personnel requirements for developing a viable standards program, Council in September, 1963, modified the Standards Committee's charge, eliminating the clause instructing it to independently develop standards. It was charged, instead, with simply monitoring and working with other standards groups.[61] Thus, through the remainder of the 1960s and 1970s most Institute activities in standardization were carried out through the ASA and its successor body, ANSI, with AIChE representatives being most active in the preparation of nuclear standards. AIChE, for example, served as the secretariat for the N46 Committee of ANSI on the nuclear fuel cycle. Meanwhile, the Standards Committee itself slowly lapsed into inactivity.[62]

The Institute's equipment testing procedures program, a "standards-like" activity initiated in 1944 to establish procedures for testing the performance characteristics of chemical engineering equipment, on the other hand, remained relatively active through most of the 1960s. Like the Institute's other standards activities, the Equipment Testing Procedures Committee relied entirely on volunteers. Its testing procedure for plate distillation columns, for instance, was produced over a period of four years and required a total of 5000 manhours, all contributed on a voluntary basis by committee members.[63] Throughout most of the late 1950s and the 1960s, it remained active under strong committee leadership, and by 1968 it had produced 14 testing procedure manuals.[64]

The committee went dormant in the late 1960s and early 1970s. In November, 1972, J. Y. Oldshue (President, 1979) reported to Council that the ETPC was "passive" and that no new procedures had been published in the previous four years.[65] Only in 1973–74 did the committee revive. By 1977–78 it was once again preparing new procedures and had begun to revise older procedures.[66]

Like its standards activities, the AIChE's research activities were reduced in the 1960s. The primary factor behind this reduction was not, as with standards, the necessity of depending almost completely on the services of volunteers, but the lack of consensus in the Institute over what the scope of its involvement in research should be.

As noted in the previous chapter, AIChE's involvement in research began with the creation of the Research Committee in 1950. This committee in the early 1950s raised funds from 36 companies for research on bubble tray efficiency in distillation columns. This research was carried out between 1952 and 1957, under AIChE sponsorship and management, at three universities, with the final report appearing in 1960.

This project created some disenchantment with research in AIChE. While it was generally conceded that the bubble tray project stimulated research in a "fruitful direction," it failed to accomplish its stated objective: developing a dependable method for predicting efficiencies.[67] Some prominent figures in the organization felt that AIChE should not sponsor research unless it could meet stated objectives, while others objected in principle to Institute involvement in research fund raising and management.[68]

On the other hand, the Research Committee was dissatisfied with the restrictions placed on its activities by Council. The society's official research policy, laid down in 1956, had restricted sponsored research to "broad and fundamental investigations which would yield results of the widest possible usefulness to the profession as a whole." Because industrial research needs were often narrow, the Research Committee had difficulty putting together a series of projects which would meet Council's requirements and yet be of sufficient value to industry to merit financial support.[69]

The difficulty of finding a project which fell within Council's guidelines and which would still attract industrial support and Council's growing doubts about the advisability of involvement in research delayed the immediate initiation of a second research project. AIChE's leadership, in the meantime, sought to redirect the committee's energies. In 1961 the Executive Committee of AIChE notified the Research Committee that it should not consider the solicitation of funds as its most important function and should act more in an advisory than in a management capacity and suggested several projects.[70] In 1964, in keeping with this charge, the Research Committee published a survey of active research projects in chemical engineering in American and Canadian universities and a compilation of suggestions for needed research.[71]

Meanwhile, a second AIChE-sponsored research project, organized under the auspices of the Machine Computation Committee instead of the Research Committee, had been launched. In April and June of 1960, shortly after developing standards for publishing computer abstracts and manuals describing computer programs, the Machine Computation Committee proposed to Council that AIChE sponsor a project to develop a set of computer subroutines for estimating the physical properties of chemical compounds and mixtures using available data and known correlations.[72] In September of 1960 Council authorized preparation of a prospectus and solicitation of research proposals from universities and research agencies and sought National Science Foundation support for the project.[73]

After a long delay, the National Science Foundation rejected AIChE's request for funds on the grounds that no basic research was involved. In May, 1962, Council, less willing than previously to become involved in fund raising for research, narrowly authorized a study of the feasibility of approaching industry for funds to support the physical properties project.[74] In August, Council authorized initiation of an industrial fund raising campaign centered in the national office.[75]

Raising funds for the computer estimation project proved more difficult than anticipated and seemed to confirm Council's doubts about AIChE involvement in research. Several companies, because of the long lag between the initial approach in 1960 and the beginning of research, had initiated computer estimation projects of their own in restricted areas and were no longer willing to contribute to the project.[76] Despite this, Council in March, 1963, approved continuation of the fund raising drive and authorized execution of a research contract with A. D. Little, Inc.[77]

The physical properties project, completed in mid 1965, was of "high technical caliber." Unfortunately there were also problems. The program was written in FORTRAN II, the most generally used scientific programming language in 1963. But by 1965 FORTRAN II was being replaced by FORTRAN IV, making the program at least partially obsolete by the time it was complete.[78] Nonetheless, the results of the project were still valuable, and AIChE offered the magnetic tapes and three volumes of documentation produced by Little for sale, leaving it largely up to buyers to maintain and update the program.[79]

EXPANDING PROGRAMS   67

In 1963 the issue of the Institute's increasing involvement in special project fund raising for research and in research management came to a head. At its March, 1963, meeting, Council seriously considered abandoning the computer estimation project.[80] At that same meeting it approved appointment of an *ad hoc* committee to review the Research Committee's charge and current AIChE research policy.[81] In May, 1963, this committee recommended that Council rescind the section of the Research Committee's charge which permitted it to initiate research programs and seek funds to support them.[82] Council formally approved a revised policy statement on research, incorporating these views, at its July, 1964, meeting.[83]

With the elimination of any active, direct role in promoting research and the restriction of their activities largely to advisory functions, the committees most involved in promoting research in the 1950s and early 1960s—the Research Committee and the Machine Computation Committee—steadily declined in activity.[84]

The same factors that had reduced AIChE commitment in standards (reliance on volunteers) and research (lack of consensus) in the 1960s also had an influence on the fate of several new programs initiated by the Institute in that period.

NEW PROGRAMS

Perhaps the most noticeable feature in the history of the American Institute of Chemical Engineers in the dozen years following its golden anniversary in 1958 was a modest but, nonetheless, symbolically significant shift toward increased activities in areas outside of the traditional ones (meetings and publications).

A number of factors, some internal, some external, contributed to this development. One of the important internal factors was the Institute's broader membership base. The executives and consultants who had dominated the early organization had, for the most part, been little concerned with things like the employment conditions of chemical engineers. They were employers and did not have to worry in the same way as employees about things like salaries, engineering oversupply, pension benefits, and job mobility. The academic, lower management, and non-supervisory engineers who entered the AIChE in large numbers between 1930 and 1960 were much more interested in such matters and wanted the Institute to increase its involvement in areas beyond the traditional ones.

External developments further contributed to the initiation of AIChE programs in non-traditional areas. In the 1960s declining enrollments in engineering, the increased role of government in funding research and development, the "space race" with the Soviet Union, the civil rights movement, the fear of obsolescence generated by the explosion of knowledge and publications in the post-World War II era, the rising importance of computers, and the changing nature of chemical engineering education were among the factors which provided stimuli for the development of new AIChE programs in areas such as minority career guidance, international relations, governmental interaction, and continuing education.

There were significant indications of increased interest in expanding Institute involvement in non-traditional fields at both the member and policymaking levels shortly after AIChE celebrated its 50th anniversary. For instance, an informal survey conducted in 1959 by an *ad hoc* Committee on Membership Services, chaired by C. A. Stokes (Director, 1956–58), revealed that members wanted the AIChE to undertake a number of novel projects: salary surveys, employment services, and surveys of corporate and management employment practices were among them.[85]

On the policy-making level the most significant indication of increased interest in non-traditional programs was the "Dynamic Objectives" report of 1961. In early 1959, as the Stokes Committee began its study of membership services, Council, influenced by concerns over the future of the profession expressed by Warren McCabe (President, 1950), appointed another *ad hoc* committee with much broader objectives. It would appraise the present status of the chemical engineering profession as a whole and attempt to determine its future and how the AIChE could best react to it.[86] This suggestion fell on fertile ground. The whole nature of chemical engineering had been changing in the post-World War II era. Traditional unit operations were being displaced by increased emphasis on the underlying sciences and on mathematics. Computers and mathematical computations had begun to play a larger role in chemical engineering. The explosion of knowledge had created a fear of technological obsolescence, and the early spectacular successes of the Soviet Union in space had cast a shadow over American engineering. Thus, in February, 1959, the Institute's Executive Committee voted unanimously to recommend that Council appoint a special committee to determine dynamic objectives for chemical engineering.[87] This committee was appointed in the spring of 1959 and asked to develop a view of the future of chemical engineering, to set objectives for the various elements of the profession (industry, education, individual engineers, and the professional society), and to determine how these objectives could be best met.[88]

Chaired by Donald Katz (President, 1959) of the University of Michigan and Robert White of the Atlantic Refining Co., the Dynamic Objectives Committee, with the aid of a grant from the National Science Foundation, attempted between 1959 and 1961 to carry out its charge. To secure outside input the committee sponsored open forums on the goals and objectives of chemical engineering at three national Institute meetings in 1959 and 1960. The committee, then, presented a preliminary draft of its report for discussion at the Washington national meeting in December, 1960. The final draft, printed in *Chemical Engineering Progress* in October, 1961, was selected for an Award of Merit by the United States Chamber of Commerce.[89]

Most of the Dynamic Objectives report of 1961 was rather conservative, endorsing traditional practices or pointing to areas rather obviously in need of refinement and improvement. For instance, it did not recommend any major changes in the organization of the Institute. The traditional goal of keeping chemical engineers informed of new developments and ideas was largely to be met by improving already tried and proven means such as an expanded divisional structure, more effective local sections, and improved programming. The report endorsed AIChE's 30-year-old efforts to broaden its membership base, declaring that it should be the voice of the entire chemical engineering profession and, as such, had "obligation to attract into membership all qualified chemical engineers." The report repeated AIChE's opposition to efforts to extend compulsory registration to all engineers and opposed the trend toward reducing the amount of chemistry in undergraduate chemical engineering curricula.

Even though the Dynamic Objectives report was generally conservative and focused on traditional means of improving the Institute's services, there were elements in the report which signaled a slight but significant shift in Institute policy—a shift toward more activity in areas outside of publications and meetings. For example, among the objectives set for the Institute by the committee were increased concern for the number of people entering the profession, the development of a post-college training program for chemical engineers, increased involvement in areas such as professional legislation and safety, and the creation of a "strong professional voice" for chemical engineers. These new goals were to be met by expanding the Institute's public relations and career guidance programs, by developing "new" educational programs, and by considering a dual organization within AIChE to permit more effective expression on professional issues without loss of the benefits of tax status as an "educational" society, among other ways. The development of a role for the Institute in employer-employee relations was not made a "dynamic objective" for the AIChE in the initial report. But, in response to comments on the preliminary draft, the Dynamic Objectives Committee added a section to its final report which provided specific suggestions to employers on the role they should play in advancing professionalism among their chemical engineering employees.[90]

Education

The shift to increased Institute involvement in areas outside of meetings and publications was most obvious between 1958 and 1970 in the field of education, specifically in the areas of career guidance and continuing education. A number of concerns, some already mentioned, contributed to the society's growing interest in educational programs. Declining enrollments in chemical engineering in the late 1950s and the widespread national belief that more engineers and scientists were needed to overcome the Soviet lead in space technology were among the factors stimulating increased AIChE activity in career guidance. Others were the rapid growth of knowledge, the rapid introduction of new techniques (like the use of computers), and a shift in college curricula to a more heavily theoretical approach to engineering. These developments had created, in the early 1960s, a fear that old chemical engineers were rapidly on their way to becoming technologically obsolete and that continuing education programs were necessary to counter this.

There are several factors which explain why the AIChE's new programs in education were more significant in the period between 1958 and 1970 than its new programs in any other area. First, far more than in any other area, besides meetings and publications, the AIChE had a tradition of involvement in education. This dated back to the foundation of the organization and had peaked with the AIChE's pioneering efforts in accrediting chemical engineering curricula in the 1920s. Second, chemical engineers generally had long placed more emphasis on the importance of education than most other engineers. In 1947, for instance, 12.7% of all chemical engineers held doctorates. By contrast only 1.8% of all civil engineers, 2.4% of all electrical engineers, and 2.0% of all mechanical and industrial engineers held doctorates. Even though, in 1947, only 9.7% of all engineers were chemical engineers, 33.8% of all doctorates in engineering were held by chemical engineers.[91] Finally, education was the one area, other than meetings and publications, on which all of the various interest groups within the AIChE—upper management, academia, lower management, non-supervisory engineers, etc.—could most easily reach agreement and from which all could benefit.

The first sign of increased AIChE activity in education came in the area of career guidance. This activity was stimulated, primarily, by a steady decline in engineering enrollment generally and in chemical engineering enrollment specifically. After a steady rise from 1951 to 1957, enrollment in engineering at American colleges peaked in 1957 at 79,000 and began to decline, slipping to 65,000 by 1962.[92] Chemical engineering shared this decline. Moreover, many educators felt that the decline in enrollment was, for chemical engineering, not only quantitative, but qualitative. Brighter high school students, once attracted to chemical engineering, were enrolling in physics, chemistry, and electrical and electronics engineering because missiles, rockets, computers, and pure scientific research had become the new glamor areas.[93]

The AIChE's Public Relations Committee by 1960 felt that this situation had become a serious threat to the chemical processing industry's future supply of manpower. In September, 1961, Stanley Adler, the committee's chairman, submitted to Council a proposal that the AIChE produce a 26-minute career guidance film directed toward students in the last years of high school and the early years of college. This film was to have two functions. It was to aid public relations by enlightening the public about what chemical engineering was and the role it played in economic development. More directly, it was to aid in career guidance by interesting teenagers in chemical engineering as a profession.[94]

Career Guidance Film

In December, 1961, Council approved in principle the idea of AIChE sponsorship of the guidance film, provided funds could be obtained.[95] By February, 1962, the Public Relations Committee had secured proposals from several film producers, had drafted a tentative letter to be used for industrial solicitation, and had developed a list of companies to be approached.[96] In May, 1962, by a nine to five vote, Council authorized the Public Relations Committee to begin soliciting funds for the guidance film.[97]

By February of 1963 the Public Relations Committee had secured most of the needed pledges. Anxious to get work underway, the committee asked Council to authorize a contract for production of the film and to provide funds to make up the difference between costs and pledges until additional funds were collected. Council approved this plan.[98] By May of 1964 the Public Relations Committee had reached its goal. More than $50,000 had been raised, roughly 87% from industry.[99] The film premiered at the 53rd National Meeting of the Institute in Pittsburgh in May, 1964, and the AIChE shortly after made the 75 copies of the film available for loan to local sections and for sale.[100]

The film circulated widely. Between September, 1964, and April, 1966, 57 copies were sold and rental copies were shown over 700 times.[101] But the impact of the film in attracting students to chemical engineering is difficult to judge, in part because enrollments began to rise in 1963 and 1964, just as the film was released. Further, the film had a relatively short life. By the late 1960s changes in dress and in employment and enrollment situations had made the film obsolete. In 1969 and 1970 the Career Guidance Committee attempted to remedy this by editing the film, shortening it by around seven minutes.[102] This extended the film's useful life. It was still being occasionally used in the late 1970s, but changing conditions had again made it obsolete, and the Institute by 1978 was discouraging its use.[103]

While the film was the national AIChE's most ambitious venture into career guidance activities, the Career Guidance Committee produced some additional aids, including several career guidance booklets (one distributed with the film), a slide show on chemical engineering careers, and a filmstrip designed to prepare seniors in chemical engineering for industrial life.[104]

The Institute's career guidance activities in the 1960s were a significant break with tradition. Prior to 1960 this work had remained on a relatively small scale and was almost entirely confined to local sections. The guidance film represented a significant national involvement in the area. Moreover, the initial purpose of the film was to attract students into chemical engineering. As Van Antwerpen noted in a *CEP* editorial in November, 1959, the AIChE had never before "beaten the drum for *more* chemical engineers."[105] It had in its first half century always depended on the supply-and-demand cycle to take care of the problem.

The continuing education program developed between 1958 and 1970 was more significant in the long run than national activities in career guidance. The background of AIChE involvement in continuing education lies in the rapid growth of knowledge after World War II, a growth encouraged by the introduction of computers, by advanced mathematical techniques, and by the heavy investment of the federal government in research. By the late 1950s the flood of new research and new techniques in almost every field of engineering had begun to create concern that older engineers were technologically obsolete and that the half-life of newly graduating engineers would be only around ten years if they were provided with no post-college studies. Concern over this problem was first explicitly broached by engineers in upper management in the late 1950s. The prospects of technological obsolescence attained national prominence when these concerns were published in widely-read media like the *Wall Street Journal* in 1961.[106] The AIChE's dynamic objectives of 1961 reflected this concern, calling for AIChE "to encourage continual improvements in education whether in formal courses or in postcollege training."[107]

The AIChE between 1958 and 1963 had begun to experiment with continuing education. In 1958 the national Program Committee offered the first of a series of "Advanced Seminars." These courses, usually given just prior to a national meeting on a reservation basis, covered highly theoretical, frontier research topics. Between 1958 and 1963 from one to three of these seminars were offered annually, with an average attendance of around 70.[108] The local sections of AIChE were somewhat more active in continuing education. A 1962 survey indicated that of the 78 sections, roughly two-thirds (52) had given some sort of refresher course during the year. These had ranged from 2-hour to 30-hour meetings.[109]

Between the extremes of new frontier information being provided on occasion by the national organization and refresher courses designed to aid the recall of forgotten information being offered by local sections, there was a broad area left nearly untouched (except by an occasional local section offering): courses to provide chemical engineers long out of college with material being currently offered in chemical engineering curricula.[110] It was in this largely untouched area that, beginning in 1963, the AIChE concentrated its efforts.

The creation of a strong AIChE program in continuing education was primarily due to W. Robert Marshall, Jr. (1963 President) of the University of Wisconsin, who made the development of such a program a theme of his year in office. In May, 1963, he first suggested to Council that it create a Committee on Continuing Education to advise the Institute.[111] In June of 1963 he brought up the idea again, this time to the Executive Committee. He was asked, however, to first review what current AIChE committees were doing in the area.[112] In November, Marshall once more brought the matter before the Executive Committee and, this time, secured approval for the creation of a Continuing Education Committee.[113] Little of course, could be done to develop a comprehensive program during his remaining month or so in office, but, subsequently, as chairman of the Continuing Education Committee, he continued to play an important role in the evolution of AIChE's program.

In 1964 the AIChE's Continuing Education and Program committees introduced the "Today Series." The program's title reflected the philosophy adopted by the Institute in continuing education: providing practicing engineers with courses in today's chemical engineering curricula. The first of the Today Series courses, "Mass Transfer Today," was offered at the 52nd National Meeting in Memphis. The continuing education program grew slowly but steadily between 1964 and 1966. In 1966 the Institute took two additional steps which significantly strengthened its program. First, Council called for the AIChE to assume leadership in providing for the continued education of its members.[114] Second, Hal Abramson was transferred from the *CEP* staff to direct efforts in continuing education.

The AIChE's education program expanded at a rapid pace. In 1966, the AIChE had offered one Advanced Seminar and four courses in the Today Series, with a total attendance of 292. Several additional types of programs were added during the next few years. In 1967 AIChE introduced Management Seminars to teach management concepts and skills to engineers already in management or moving into management. In 1968 the Advanced Seminar Subcommittee of the national Program Committee was merged into the Continuing Education Committee. And in 1969–70 the AIChE expanded its continuing education program to include workshops where people with common problems could meet in relatively small groups for a program of papers and problem discussion. In 1970 the AIChE offerings in continuing education included three Advanced Seminars, 10 Management Seminars, one Workshop, and 34 courses in the Today Series: a total of 48 courses in all, with a total attendance of 1285.[115] Continuing education had by this time become one of AIChE's most significant activities, occupying a considerable proportion of the program space at national meetings, and the American Institute of Chemical Engineers had become a leader in the field.

The AIChE's involvement in more traditional collegiate education also increased in the 1950s and 1960s, largely prompted by concern over the rising popularity of common pre-engineering programs and the spread of engineering programs without the traditional adjectival designations (i.e., chemical, civil, mechanical, etc.).

The Common Core Controversy

The strong position of chemistry in chemical engineering programs has traditionally distinguished chemical engineers from engineers in other fields. AIChE's leadership, as well as many of its members, feared that the spread of common core curricula and general engineering programs, both of which minimized chemistry requirements, could destroy chemical engineering as a distinct and important engineering discipline. J. L. Franklin, then chairman of the Education & Accreditation Committee, explicitly expressed such fears in his 1963 report to Council:

> More serious for chemical engineering, of course, will be the problem of including enough chemistry in the [common core] undergraduate program to enable the pre-engineering students to specialize in chemical engineering at the graduate level. Unless this is done the profession of chemical engineering will either disappear or it will be necessary for chemical engineering to return to close association with chemistry and withdraw from the engineering colleges.[116]

EXPANDING PROGRAMS

Franklin's fears were reiterated in 1967 by Past President S. W. Churchill:

> Chemical engineers are different from other engineers because of their roots in the science of chemistry. If all engineering were forced into a single mold, chemical engineering would be watered down or eliminated. This has indeed happened in most institutions which have adopted generalized engineering or gone very far in the direction of a common core since the amount of chemistry in the curriculum is invariably decreased.[117]

The fear that the very existence of chemical engineering as a major engineering discipline was being threatened accounts for the Institute's zealousness in guarding its special position in ECPD and its militant opposition to plans to extend common curricula in engineering proposed by an American Society for Engineering Education (ASEE) committee.

In 1961 the Engineers' Council for Professional Development asked ASEE to conduct a study on engineering education to guide future policy decisions in the education and accreditation area. The ASEE committee charged with making the study secured a grant from the National Science Foundation, and, in October, 1965, after several years of study, issued a preliminary report titled "Goals for Engineering Education." This was followed by an "Interim Report," delivered in September, 1967, and a "Final Report," delivered in January, 1968. Three of the recommendations of the Goals Study Committee were particularly objectionable to AIChE: (1) that the first professional degree in engineering should be the master's degree, awarded after five years and identified as a Master of Engineering Degree without qualifying adjectives; (2) that the four-year Bachelor's Degree continue to be offered but only as an introductory engineering degree; and (3) that accreditation by ECPD be changed from specific curricular accreditation to accreditation by overall engineering unit (i.e., by college of engineering as a whole, instead of by department).[118]

Questions of curricular revision are normally the concern of educators. Thus from the 1930s on the AIChE had not taken strong or vocal stands in the area of education, restricting its input largely to periodic accreditation visits. But because the growth of common core programs which excluded or severely reduced chemistry requirements, the Institute's leadership viewed the proposals of the ASEE's Goals Committee as a very serious threat. Breaking with the tradition of the previous two or three decades, the Institute publicly and vocally opposed the ASEE proposals. Council issued position papers critical of the "Preliminary Goals Report" in March, 1966; critical of the "Interim Report" in August, 1967; and critical of the "Final Report" in December, 1968.[119] In addition, the Institute added a special evening session to the agenda of its Philadelphia meeting in December, 1965, to discuss the report and held an open forum on the report at its Salt Lake City meeting in the spring of 1967. *CEP* had extensive coverage of the controversy.[120]

The Institute's opposition to the segments of the "Goals Report" mentioned above undoubtedly had its roots in the organization's early struggles to establish the legitimacy of chemical engineering as a distinct discipline and in fears that the adoption of the report's recommendations would destroy that distinction. In the position paper issued on the preliminary report, for instance, Council noted, in response to the call for institutional rather than program accreditation, that the AIChE "does not presume to pass upon the merits of staff and facilities for instruction in other branches of engineering, and by the same token does not intend to relinquish its responsibility for setting standards in its own field." Against the call for undifferentiated curricula leading to a general degree in engineering at the end of the fifth year, the Institute's ad hoc committee on the "Goals" argued that specialized education in chemical engineering was necessary for producing engineers with the skills needed for the chemical processing area.[121]

The AIChE's leadership was strongly backed in its public opposition to the Goals Report by practically all segments of membership. The Ohio Valley and Detroit sections adopted resolutions condemning the report as "inimical" to the interests of the AIChE, the chemical engineering profession, and industry.[122] The Interim Report was discussed at the 1967 Salt Lake City meeting of the Institute, as already noted.[123] This meeting drew 100 "hostile" participants. One reporter noted that in the two-hour session no one made a comment favorable to the report, despite repeated pleas from the chairman for someone to speak up in support of it.[124] Participants at the meeting were asked whether the AIChE should ignore the report, put out a mild general statement opposing the report, or "express its objection strongly and vocally." With one exception (a vote to ignore the whole issue), those present voted to "violently" oppose the report.[125]

At least partially as a result of the "unremitting hostility" of AIChE to critical sections of the Goals Report, the ECPD, on receiving the final report in October, 1968, accepted it merely as input for future plans and chose not to attempt to implement its recommendations.[126]

The strong opposition within the chemical engineering profession to proposals posing a threat to the existence of chemical engineering as a distinct discipline may also help explain the AIChE's position on an ambitious attempt made by the Engineers' Joint Council (EJC) to unify the engineering profession in the late 1960s.

From its foundation in the early 1940s to the mid 1960s the EJC had been a loose federation of engineering societies, serving as a forum for discussing possible areas of cooperation. In 1967, in an effort to strengthen its program, it adopted a new constitution. The EJC had traditionally had only a single class of members: engineering societies. The new constitution provided for industrial and individual membership as well. Management had traditionally been vested in a board made up of representatives appointed by the constituent societies and largely controlled by the Founder Societies. The new constitution provided for a management board elected by an assembly whose members were encouraged to act as individuals without a tie to the engineering societies. Essentially, the new constitution attempted to create an organization, independent of the organized engineering societies, with a claim to represent the entire engineering profession.

AIChE's leadership objected to the new constitution on broad and fundamental grounds. They objected to the non-democratic election procedures proposed for the new organization, to the reduced influence of the major American engineering societies, and to the creation of a body that could take public positions without the consent of the major engineering societies. At the most basic level, however, many members of AIChE's ruling structure opposed a merger of all engineers into one large, centralized society for much the same reason they opposed the merger of all engineering curricula into a common core: such a merger posed a threat to the viability of chemical engineering as an independent profession.

When the Engineers' Joint Council adopted the new constitution over AIChE's objections in 1967, the Institute withdrew from the organization effective January 1, 1968.[127] To ease the impact of its withdrawal, however, the Institute, at the suggestion of

George Holbrook, then serving as Treasurer, approved payment of three months additional dues to EJC.[128] Only in 1973, after the EJC had again modified its constitution, eliminating many of the objectionable features and giving member societies more power, did AIChE rejoin.[129]

Other Non-Traditional Programs

In the 1960s the Institute's new programs in other non-traditional areas like minority career guidance, pensions, governmental interaction, and international relations were much more significant than previously, but they did not enjoy the same degree of success as its new programs and initiatives in education or its traditional meetings and publications programs. Two factors largely account for this divergent performance: lack of consensus and dependence on volunteers. There was widespread consensus on AIChE's role as a sponsor of meetings, a publisher of literature, a promoter of continuing education, a spokesman for the profession in educational matters, and even as a spokesman for the profession in intersociety affairs. There was no such agreement about what the Institute's role should be in other fields and how these roles should be played out. This made a significant increase in the level of involvement in professional, political, economic, and social affairs difficult. Moreover, AIChE, like other professional engineering societies, was dependent on the volunteered time and efforts of its members to carry out such activities. It is difficult to sustain programs with volunteers even where there is enthusiasm and widespread support. Where neither exists, programs dependent on volunteers are inevitably restricted in scope and effectiveness.

The programs initiated by the Institute in the areas of ethics and confidentiality clauses illustrate how the organization's actions were limited by the difficulty of securing consensus within the Institute. The activities of AIChE in the areas of minority career guidance, governmental interaction, and international relations illustrate how the society's response to the social, professional, and political concerns of the 1960s were further limited by dependence on volunteerism.

Codes of ethics have traditionally not been given a high priority by engineering societies; they are often adopted as window dressing and then ignored.[130] Such was the case with AIChE's code. Adopted in 1912, it was primarily used to screen applications during the years when AIChE was a small, elite society, but was little used after 1930. In 1947 the Engineers' Council for Professional Development, in cooperation with its constituent bodies, including AIChE, produced a set of "Canons of Ethics" to replace the multiplicity of engineering ethics codes developed by the multiplicity of engineering societies. AIChE "assented" to the ECPD canons, but retained its own code.[131] In 1961 AIChE's Executive Committee suggested that ECPD rewrite its canons and submit the revision to AIChE for possible full adoption.[132] The revised canons, completed in 1963, were adopted by AIChE in 1966 to replace its 1912 code.[133] At about the same time, the Institute voted to move the code of ethics from the AIChE constitution to its bylaws.

In August, 1965, AIChE President W. B. Franklin noted that local sections were becoming "quite critical" of the national organization "owing to the lack of effective Institute action" in the area of ethics and urged the Professional Development Committee to report back to Council as quickly as possible.[134] This led to a series of discussions in late 1965 between Council and the committee on the issue of an ethics advisory service and the

Institute's potential role as a "friend of the court" in cases involving ethics.[135] By early 1966 both groups had concluded that the AIChE should stay out of the ethics advising business and out of individual court cases.[136]

Why did the Institute refrain from initiating major new programs in the ethics field? The key was probably, as noted, lack of consensus. On specific issues there was often no wide agreement between the Institute's major interest groups on what was right and what was wrong.[137] Recognizing this, the organization's leadership backed off from involvement in individual ethics cases, preferring to promote only the general principles on which there was agreement and to provide forums so that opposing sides in ethics matters could exchange views.[138]

For similar reasons Council was also cautious and conservative on a related professional question—confidentiality clauses, the agreements which engineers are usually asked to sign when joining a company. Confidentially clauses forbid engineers to divulge proprietary or confidential information detrimental to their companies both while employed by them and after leaving for other employment. These clauses are a matter of significant concern to engineers because engineering is a highly mobile vocation, and much of this mobility is dependent on transferring skills and knowledge from one job to the next. Seven years after graduation less than half of all engineers were still with their original employers in the 1960s.[139]

Between 1967 and 1969 the Professional Development Committee studied confidentiality clauses, surveying 52 firms in the chemical processing industry on their practices. Using these data the committee wrote a "model confidentiality clause" which it considered fair to both employer and employee. It bound the engineering employee to protect the proprietary material of past employers, but limited the time period that most material could be protected and made specific identification of material considered confidential a responsibility of the employer.[140]

The model confidentiality clause was, like the ethics advisory board, a potentially divisive issue. Engineers in upper management feared that the model clause would be used by job-hopping engineers to carry proprietary knowledge from job to job. Younger engineers feared that the model clause could be interpreted in a manner that sharply restricted transferable skills and experience. Moreover, some prominent members of AIChE had serious reservations about "interjecting" the AIChE between a member and his employer."[141] The motion to adopt the model clause failed on an eight to eight vote in the May, 1969, Council meeting.[142]

Several Council members who had voted against adopting the model clause, however, believed that the Institute needed to take some action, albeit a more conservative one. Thus, in November, 1970, Council approved a set of guidelines which reviewed the issues involved in the trade secret/confidentiality clause conflict and laid down general principles to guide employers and employees confronted with problems.[143] The approach here was analogous to the position in the ethics area. It laid down general principles where broad consensus could be attained; it did not dictate specifically how the clauses should read or review specific cases.

Pension Concerns

The engineering profession's high job mobility made pensions, as well as confidentiality clauses, a matter of serious concern to chemical engineers. In the 1960s most company pension plans had long vesting periods, that is, most companies required employees to remain with them from 10 to 20 years before they acquired non-forfeitable rights to the company's contributions to their pensions. A chemical engineer who changed jobs every 8 to 12 years could easily find himself at retirement with no pension rights save social security. An American Chemical Society survey in the 1960s indicated that as many as 50% of all American chemists and chemical engineers might have practically no pension on retirement.[144]

The long vesting periods required by most company pension plans and the absence of portability in pension plans (i.e., the ability to move pension savings from employer to employer) were viewed by non-supervisory and lower management chemical engineers as attempts by upper management to restrict job mobility and tie employees to unsatisfactory jobs. Employers, on the other hand, viewed the long vesting periods as a method of holding experienced people, minimizing loss of know-how to rival firms, and minimizing out-of-pocket expenses for pension plans.[145]

Several surveys in the mid 1960s indicated growing interest in the problems of vesting and portability in pensions. A 1964 ACS survey, for instance, indicated that 85% of its members favored ACS development of an industry-wide system for transferring pension rights of members who changed employers.[146] Moreover, almost 80% of chemical engineers responding to a 1964 questionnaire favored engineering societies taking action to establish some type of portable pension plan.[147]

Due to this interest, 1965 AIChE President W. B. Franklin informally approached the Professional Development Committee, asking it to look at the idea of portable pensions and what AIChE might be able to do. Richard Bergman was placed in charge of the subcommittee to investigate the area.[148] The whole question proved highly controversial, provoking "spirited" discussion when Bergman presented his preliminary report to the Professional Development Committee in December, 1965.[149] A request to local sections for feedback on the issue also yielded a "certain diversity of opinion." Most felt the Institute should do something in the pension area, particularly in conjunction with other societies. But some felt that pensions were not a "suitable topic for the Institute to even discuss." And even those who favored action did not think that the Institute should contact companies about the issue directly.[150]

After the issue of portable pensions had been discussed at various meetings the Professional Development Committee asked Council to approve a general position statement which declared that "early vesting and portability" would contribute both to "professional development of chemical engineers, and to the continued growth and viability of the national economy." Because of the lack of consensus among the major interest groups in the organization on the issue, Council, instead, approved a modified statement which simply declared that the Institute was studying and would continue to study the possible effects of early vesting and appointed an *ad hoc* Committee on Vesting and Portability of Pensions.[151]

This committee presented its report in September, 1967. It outlined the pros and cons of early vesting and portable pensions, recommending that the AIChE favor both since they promoted a free market for engineers. Because the petroleum and chemical industries already had established retirement plans with reasonably early vesting and the academic world already had portability through the Teachers Insurance Annuity Assoc. program, the committee recommended no action in these areas. But it did recommend that the Institute consider establishing a retirement program for chemical engineers who were self-employed or employed by companies with no retirement plans and that this action be considered in conjunction with the American Chemical Society, which had already undertaken a detailed study of portable pensions.[152]

Council accepted the latter recommendation, authorizing the appointment of a liaison with the ACS to keep the Institute informed of that society's investigation and possible participation in any plans it developed.[153] The decision to merge AIChE's pension interests with the development of ACS's plan effectively ended Institute action in the pension area for the decade of the 1960s.

The absence of widespread consensus among the various groups which composed AIChE (upper management, middle management, lower management, non-supervisory engineers, academic engineers) on the role the organization should take in areas like ethics, confidentiality clauses, and pensions clearly placed limits on what could be done in these areas.[154] In several other fields in which the Institute initiated activities, its programs were also limited by having to depend almost entirely on volunteers, since the AIChE, like most engineering societies, did not have the financial resources or membership commitment to those programs to do otherwise. Examples of such programs were minority career guidance, governmental interaction, and international relations.

Minority Career Guidance

The civil rights movement and the rising consciousness of the plight of the American Black were among the major social phenomena of the 1960s. The chemical engineering profession was not left untouched by civil rights activism. Due to decades of discrimination in education and the absence of chemical engineering programs in the few predominantly Black colleges, Blacks were scarce in all branches of engineering and scarcest of all in chemical engineering. In a period of engineering undersupply, the hope of filling engineering needs from the ranks of minority groups, while at the same time attempting to remedy past neglect, provided the stimulus for the initiation of an Institute program in minority career guidance.

The initiative for AIChE's program came from its New Jersey Section. In the fall of 1966 the New Jersey Section attempted, with the help of AIChE National, to secure a grant from the Ford Foundation for a national minority youth career guidance effort for chemical engineering. Then in March, 1967, following an address by a Black active in minority youth motivation, the New Jersey Section formed an *ad hoc* committee headed by Gerald A. Lessells (Director, 1973-75) to consider further actions. That summer the section ran several career guidance sessions with Black youth in the region. It then attempted to interest the national organization in encouraging similar programs in other local sections.

In December, 1967, Lessells was named chairman of a national task force on minority youth career guidance. In May, 1968, the task force asked Council to make minority youth career guidance a permanent activity of the AIChE. It recommended that the society adopt a position statement supporting minority career guidance efforts, put together a package of materials on minority career guidance, and aggressively seek industrial support in the form of scholarship aid and summer jobs for disadvantaged youth.[155] Council accepted these recommendations and asked the task force to develop an implementation program.[156]

The program, adopted in September, 1968, put the burden of implementation on local sections since they were best equipped to approach local industry, establish ties with local minority groups, and work through ghetto schools. It also placed considerable emphasis on involving minority technical personnel in the effort.[157] The new program was publicized in *CEP* and promoted by presidential letter and Council liaison.[158]

During the next few months the Task Force for Disadvantaged and Underprivileged Youth, now a permanent subcommittee of the Career Guidance Committee, held sessions at two AIChE meetings. National AIChE headquarters, in the meantime, sent minority career guidance packets to 61 local sections.[159] Despite these efforts, the results were disappointing. In May, 1969, Council liaison to the Career Guidance Committee, Director T. W. Tomkowit (President, 1972), reported that "little progress" had been made in interesting most local sections in the program.[160] It was estimated that only 4 of 91 local sections had established active programs.[161]

In the summer of 1969, in an attempt to increase interest, AIChE President H. D. Guthrie sent a joint letter with the president of the American Chemical Society to the local sections of both organizations, urging joint action on minority career guidance. But this had no great effect.[162] In August, Tomkowit informed Council that the subcommittee had "gone as far as it can in terms of interesting local sections." Three local sections had "substantial active programs" (New York City, New Jersey, Sacramento), five had "some but not substantial activities within 'conventional' career guidance," and the remainder of the local sections had no activity.[163] In subsequent months only a few more local sections developed programs (e.g., St. Louis, Delaware Valley).[164]

The Institute's attempt to expand its counseling activities into minority career guidance on a major scale in the 1960s and early 1970s was limited by factors largely beyond its control. Lack of enthusiasm for such a program was clearly evident in a large number of local sections. In addition, the problems faced by the few sections which embraced the program were immense. Poor study habits, lack of motivation, and the absence of basic skills in language and mathematics were things that could not easily be remedied by a handful of dedicated, part-time volunteers. Thus even in sections where enthusiasm was initially high (for example, St. Louis),[165] interest waned as the immensity of the problem became obvious.

Governmental Interaction

Another new activity undertaken by the AIChE in the 1960s which remained limited in scope due to apathy among membership, the immensity of the task, and the necessity of depending on voluntary help was governmental interaction. AIChE interest in closer ties with government was stimulated by several factors. One was the enormous increase in the federal government's involvement in technological affairs in the post-1945 era. This increased involvement came in several areas. After World War II, for instance, the federal government became increasingly involved in funding technological and scientific research and development. In 1964 it spent nearly $15 billion, 15% of the total national budget, on research and development, more than industry and the universities combined. Another factor was the growth of government regulatory activities in the chemical process industries, particularly the imposition of much stricter controls over air and water pollution. An additional stimulus to action was the engineering community's belief that federal technological programs were too heavily influenced by scientists and too little influenced by engineers because scientists had traditionally been more strongly involved in political affairs than engineers.

Because of increased concern over government's role in technology, F. W. Hammesfahr, the AIChE's representative on the Engineers' Joint Council's government liaison committee, asked Council in May, 1964, to form an *ad hoc* Committee on Governmental Relations.[166] This committee recommended in November, 1964, that Council create a permanent government relations committee and "not be afraid of controversy."[167] In December, 1964, Council accepted the recommendation and in February, 1965, formally established a Government Programs Advisory Committee (GPAC), charging it with seeking areas of governmental activity where chemical engineers could provide input, preparing programs for working with and through governmental agencies, and advising Council on governmental affairs.[168]

While the creation of a permanent committee to involve the Institute with the federal government seemed to portend a significant expansion of activities in the area of governmental interaction, this did not occur. The new committee was not vocal on public issues, and AIChE's leadership was similarly hesitant to depart from the low profile it had maintained in the political arena for some decades. Reflecting this hesitation, Council in May, 1969, approved a new statement of policies and procedures for GPAC. The new policy statement limited the committee's charge and emphasized that it was *not* the policy of the AIChE to influence legislation in an organized, ongoing manner, with the possible exception of legislation of concern to *all* chemical engineers, like chemical engineering education or professional registration. The 1969 policy statement simply encouraged Institute members to be active in public policy matters as individuals or informal aggregates of individuals.[169]

Although the AIChE clearly increased its involvement in public policy matters in the 1960s, governmental interaction did not become a major program. One of the primary reasons was the recognition that political issues tended to divide engineers rather than unite them. In a *CEP* editorial, Ted Burtis, 1967 President, pointed out that the AIChE was made up of 30,000 members with a common bond of interest in chemical engineering, but with otherwise diverse views on any subject. Thus the Institute's Council, made up of 17 individuals and far from homogeneous itself, could not confidently express the views of 30,000 members on any subject. He pointed out, further, that any policy statements by AIChE outside of its own narrowly technical area would be unlikely to carry much authority and would simply add to the clamor.[170]

Finally, reliance on volunteers restricted the level of Institute interaction with government in the 1960s. The 1969 policy change which reduced the Government Programs Advisory Committee's charge, limiting the committee to keeping abreast of legislation affecting engineering, developing working relationships with similar committees of other societies, and arranging for a chemical engineering input to various legislative and executive agencies was a recognition that the initial charge placed on the committee was simply too heavy. But even the more restricted charge was too heavy. Providing effective advice to the enormous governmental bureaucracy that had emerged in the United States by the 1960s was a very difficult task. The dedicated, voluntary efforts of GPAC members were simply insufficient. In 1973 the chairman of the committee, J. C. Dart, informed Council that because GPAC was a committee of volunteers, it "did not have the time and

EXPANDING PROGRAMS 81

resources to make adequate contact to achieve even a modestly effective level of AIChE involvement in Federal Government Affairs."[171]

International Relations

Another area where reliance on volunteers to tackle major problems limited the scope of AIChE activities was international relations. The Institute's growing concern with international relations was stimulated by several things. First, international relations were a major social issue of the 1960s. The economic plight of the underdeveloped world was a source of serious concern to many and led to the creation of programs like the Peace Corps and the Alliance for Progress. Second, many chemical corporations had become or were in the process of becoming multi-national by the 1960s. Third, and perhaps most important, as chemical engineering emerged as a profession in other countries, their chemical engineers looked to the AIChE, as the oldest and largest chemical engineering society, for guidance in establishing their own organizations or for institutional affiliation.

The foundations of AIChE's attempt to develop a program in international relations were laid at a joint meeting held in June, 1960, with the Instituto Mexicano de Ingenieros Quimicos (IMIQ), an organization the AIChE had helped found. At that meeting, F. M. Tiller, Dean of Engineering at the University of Houston, suggested that the Institute establish a permanent International Relations Committee to enable AIChE to play a more powerful role on the international scene.[172] The suggestion was taken up by Jerry McAfee (President, 1960). The subject was discussed at the AIChE Executive Committee's meeting in January, 1961, and Council appointed an *ad hoc* committee to review the issue at its meeting of February, 1961.[173]

In August, 1962, D. A. Dahlstrom (President, 1964), then chairman of the *ad hoc* committee, presented its final report to Council. The committee noted that the stature of the AIChE as the oldest and largest chemical engineering society in the world was very high and that the organization's publications and programs carried high prestige. The committee argued that AIChE should attempt to capitalize on this prestige, especially since technical contacts between countries were increasing in importance. The committee outlined a number of possible projects for the proposed International Relations Committee. Among them were reception of visiting chemical engineers, liaison with foreign chemical engineering groups, provision of aid and advice in the formation of engineering organizations in underdeveloped countries, establishment of permanent ties with these groups, participation in international symposia and meetings, promotion of programming devoted to international subjects in both the national organization and local sections, and assistance to foreign engineering education efforts. Council approved formation of the new committee.[174]

The new International Relations Committee was successful in several areas such as increasing Institute participation in international meetings. In the 1960s annual meetings with foreign engineering societies or international engineering confederations became more the rule than the exception. Further, at least in part because of the efforts of the committee, the AIChE increased its participation in international confederations, joining, among other groups, the Union Panamericana de Asociaciones de Ingenieros (UPADI), the Conference of Engineering Societies of Western Europe and the United States

(EUSEC), and the Inter-American Confederation of Chemical Engineering.[175] AIChE was a founding member of the latter group, and AIChE Executive Secretary Van Antwerpen was its first Secretary-General and later, 1977, became President.[176] Finally, the International Relations Committee formulated a program of reciprocity with foreign chemical engineering societies and encouraged local sections to form their own international relations committees.

By August, 1965, 29 local sections had formed such groups.[177] Among the most successful was the South Texas Section's committee. Its international relations program included inviting foreign students to monthly meetings of the section and to members' homes, arranging inspection trips and hospitality for foreign chemical engineers, meeting jointly with IMIQ's Monterrey, Mexico, section (a practice continued annually by AIChE's Coastal Bend Section), devoting meetings to foreign affairs, and collecting books to forward to universities in Brazil, Ecuador, and Colombia.[178]

Despite some success, however, by 1972 the International Relations Committee had recognized that its program was too ambitious for a group of volunteers. An ad hoc committee appointed to study this problem reported in November, 1972, that only around ten local sections had the interest and resources to support effective action in international relations. The original objectives of the International Relations Committee, the ad hoc committee concluded, had been "very broad and general." This had been one of the committee's problems. "Past committees have tried to pursue too many areas and have been overly ambitious." Attempts to establish contact with all chemical engineering societies, to develop reciprocity with all of these organizations, to provide committee experts to all parts of the world, and to develop international relations activities in most sections were just too much to expect of a committee of volunteers. Accordingly, the ad hoc committee asked that the International Relations Committee be given more limited objectives: injecting the international perspective into programming, maintaining liaison with international confederations of engineering organizations, hosting international guests at meetings, and keeping the AIChE informed of international trends.[179]

AIChE's Council was apparently not persuaded that these activities merited a separate committee and in June, 1973, dissolved the International Relations Committee, establishing in its place an international relations area in the national Program Committee.[180] Only in 1980 did the Institute revive an International Activities Committee. It did previously take a specific action in April, 1970, that whenever possible, meetings should be arranged with international regional organizations to avoid a proliferation of bilateral ones.[181]

In summary, during the late 1950s and early 1960s the AIChE's level of activity in non-traditional fields such as pensions, confidentiality clauses, ethics, minority career guidance, international relations, and certain areas of education increased. This increase was most significant in education, where the AIChE produced a major career guidance film, developed an extensive and nearly self-supporting continuing education program, and played a significant role in killing proposals to modify the nature of engineering education. In other non-traditional fields the level of activity was clearly higher than it had been in earlier decades, but remained severely restricted in scope and effectiveness due to the magnitude of the problems involved, the lack of enthusiasm of engineers for most non-technical programs, dependence on volunteers, and the necessity of compromising and limiting activism in fields where consensus could not be attained.

CONCLUSION

The American Institute of Chemical Engineers enjoyed substantial growth in the period between 1958 and 1970. Its membership increased from 18,000 to more than 38,000. More significantly the Institute continued to attract a steadily larger proportion of American chemical engineers. Gloria Lambson (now Staff Director, AIChE Member Activities & Services) estimated that about 47% of all American chemical engineers belonged to the AIChE in 1958, a relatively respectable figure since organizations like ASCE and ASME could claim only around 30 to 35% of, respectively, American civil and mechanical engineers. By 1970, according to Lambson, the Institute's proportion of all American chemical engineers had risen to over 69%.[182] Moreover, the Institute had continued to reduce the hold of the rival American Chemical Society on chemical engineers. In the early 1950s more than half of all AIChE members were also members of the ACS. By the mid-1960s this figure had been reduced to only about one-third and was to continue to drop, falling below 10% by the early 1980s.[183]

Why did the Institute continue to grow at a relatively rapid rate in the 1960s? Partly it was due to external factors. The chemical industry remained an important growth industry in the 1960s, so the number of available chemical engineers continued to grow steadily. The Institute, however, was able to attract a larger portion of these to membership through more aggressive membership campaigns (particularly in the late 1950s and early 1960s) and through the elimination of barriers to membership. In 1958, for example, the AIChE initiated automatic transfer of Student members to the Associate member class on graduation, and in 1967 it eliminated the entrance fees previously required for membership.

These actions had an impact primarily on the recruitment of graduating seniors and can be judged by comparing the number of new AIChE members annually to the size of graduating classes in chemical engineering. In the 1948-50 period new members typically numbered only around 30% of annual chemical engineering graduates. In the period 1952-60 the proportion increased to around 60%. In the early 1960s, however, new members annually numbered more than 90% of the size of the annual graduating classes in chemical engineering.[184]

The factors outlined earlier in this chapter also contributed to the Institute's ability to attract and retain members in the late 1950s and in the 1960s. In this period, as noted, the AIChE made significant refinements in its basic administrative structure. It improved the efficiency of services by altering the traditional relationship between Council and headquarters and by adding key full-time professionals to headquarters staff. It strengthened ties to its local sections and created divisions to meet the growing needs of specialists. At the same time it improved and expanded its publications and meetings programs (including continuing education) and attempted, through the initiation of non-traditional programs, to adjust to changing social and political conditions. While the AIChE's non-traditional programs often had only limited success, they provided an outlet for the energies of members concerned about external professional and social problems and, at the least, demonstrated the organization's concern about these problems.

The priority which the Institute continued to place on its traditional publications and meetings programs raised few questions, for the 1960s were prosperous and in prosperous times most chemical engineers asked little more of the Institute than efficient operation and timely and useful publications and meetings. The Institute's professional (ethics, confidentiality clauses, pensions, governmental relations) and social (minority career guidance) programs were regarded by all but a small proportion of membership as nice, but peripheral to the central function of the Institute: the promotion and diffusion of knowledge. But, as we shall see in the next chapter, when prosperity faded, demands for a reconsideration of AIChE priorities were to increase substantially.

Chapter 3: Notes

1. *CEP*, v. 54 (August, 1958) p. 98.
2. *CEP*, v. 55 (August, 1959) p. 152; (December, 1959) p. 138.
3. An attempt to expand *CEP* coverage of candidates to include not only biographical information, but also candidate statements on program was defeated by Council. See Council, *Minutes,* November 30–December 1, 1963, p. 548; January 10–11, 1964, p. 1.
4. Council, *Minutes,* August 23, 1969, pp. 298–304. The quotes are from pages 301 ("essentially a must") and 299 ("not in the best interests").
5. Council, *Minutes,* November 15, 1969, p. 438 (Appendix G).
6. Council, *Minutes,* August 25, 1962, p. 184.
7. Council, *Minutes,* September 28, 1963, pp. 338, 430–33, and Appendix C; Executive Committee, *Minutes,* November 2, 1963, pp. 491–96.
8. Council, *Minutes,* December 5, 1964, pp. 451–52, 469–76 (Appendix F).
9. Executive Committee, *Minutes,* August 20–21, 1965, p. 293.
10. Marx Isaacs, "A Short History of the AIChE Petroleum and Petrochemicals Division, " 10-page, unpublished manuscript, AIChE files.
11. *CEP*, v. 57 (October, 1961) p. 97.
12. Council, *Minutes,* February 25, 1961, p. 20.
13. Council, *Minutes,* September 24, 1960, p. 241.
14. Council, *Minutes,* September 23, 1961, p. 274; December 2, 1961, p. 402.
15. Council, *Minutes,* January 13, 1956, pp. 10–11.
16. Executive Committee, *Minutes,* April 18, 1959, pp. 257–63. The problem of non-national members had been a matter of concern for some years, see Van Antwerpen and Fourdrinier, *High Lights,* p. 106.
17. Council, *Minutes,* December 3, 1960, p. 337.
18. Council, *Minutes,* February 25, 1961, pp. 20–22.
19. Council, *Minutes,* September 23, 1961, p. 274.
20. Executive Committee, *Minutes,* May 6, 1962, p. 73; Council, *Minutes,* May 19, 1962, p. 90f.
21. *CEP*, v. 58 (January, 1962) p. 126; (October, 1962) p. 164; and (December, 1962) p. 142.
22. *CEP*, v. 59 (May, 1963) p. 94; Council, *Minutes,* December 1, 1962, p. 393.
23. *CEP*, v. 59 (May, 1963) pp. 94, 96; Council, *Minutes,* May 4–5, 1963, p. 147.
24. A report of the Subcommittee on Local Section Activities in 1960 indicated that all of the other Founder Societies had rebate programs. See Council, *Minutes,* September 24, 1960, pp. 242–45.
25. Council, *Minutes,* November 9 and December 16, 1945.
26. Council, *Minutes,* November 30–December 1, 1963, p. 617; see also Council, *Minutes,* July 18, 1964, p. 351, and Executive Committee, *Minutes,* November 14, 1964, p. 412.
27. Council, *Minutes,* November 30–December 1, 1963, pp. 676–85.
28. Council, *Minutes,* July 18, 1964, p. 350.
29. Council, *Minutes,* July 18, 1964, pp. 350–54, for the report of the Franklin Committee. Also, Executive Committee, *Minutes,* November 14, 1964, pp. 412–13, and Council, *Minutes,* December 5, 1964, pp. 462–63, 525–26. The *Minutes* do not indicate whether any charters were actually revoked.

EXPANDING PROGRAMS   85

30. The data are from reports made by Harold I. Abramson to Council and dated May 16, 1970, and February 27, 1982.
31. Council, *Minutes,* February 3, 1962, p. 12.
32. Executive Committee, *Minutes,* May 6, 1962, pp. 72, 81–83; Council, *Minutes,* May 19, 1962, pp. 92–93, 100–05.
33. Council, *Minutes,* May 19, 1962, pp. 92–93; August 25, 1962, p. 181.
34. Council, *Minutes,* May 16, 1964, p. 239.
35. Executive Committee, *Minutes,* January 16, 1965, p. 2.
36. Council, *Minutes,* February 6, 1965, p. 26.
37. Executive Committee, *Minutes,* January 16, 1965, p. 2.
38. *CEP,* v. 53 (November, 1957) p. 208; (December, 1957) p. 130.
39. "Report of the Ad Hoc Committee on the Future of the Institute," February, 1967, in Executive Committee, *Minutes,* April 14, 1967, p. 155.
40. *CEP,* v. 66 (April, 1970) p. 7.
41. Data provided by Robert Guertin, AIChE staff.
42. Council, *Minutes,* September 28, 1963, pp. 430–33.
43. Executive Committee, *Minutes,* January 10–11, 1964, pp. 2–4; Council, *Minutes,* February 1, 1964, p. 23.
44. Council, *Minutes,* October 23, 1959, p. 481.
45. *CEP,* v. 57 (April, 1961) pp. 93–96.
46. *CEP,* v. 57 (April, 1961) p. 43 (for attendance); Executive Committee, *Minutes,* August 3, 1962, p. 163 (for profit).
47. *CEP,* v. 63 (April, 1967) p. 33.
48. *CEP,* v. 69 (May, 1973) p. 3.
49. Council, *Minutes,* November 30–December 1, 1963, pp. 622–24.
50. Council, *Minutes,* September 7, 1975, pp. 275 (the 79% figure), 276 (on Resen).
51. Council, *Minutes,* November 18, 1958, pp. 291–94.
52. *CEP,* v. 55 (May, 1959) p. 94.
53. The system is described in Rollin Morse, "Information Retrieval," *CEP,* v. 57 (May, 1961) pp. 55–58. See also B. E. Holm, "Information Retrieval—a Solution," *CEP,* v. 57 (June, 1961) pp. 73–78.
54. Saul Herner and W. F. Johanningsmeir, "Information Storage/Retrieval: Is It Working?," *CEP,* v. 61 (March, 1965) p. 25.
55. *CEP,* v. 58 (February, 1962) pp. 118–21; v. 60 (September, 1964) pp. 29–30.
56. Council, *Minutes,* June 1, 1974, pp. 275, 278.
57. Personal communication, Trevor Evans to J. Charles Forman, October 18, 1982.
58. Executive Committee, *Minutes,* February 16, 1963, p. 3.
59. Council, *Minutes,* September 7, 1975, pp. 257, 277–78.
60. Council, *Minutes,* March 9–10, 1963, pp. 114–16; September 28, 1963, pp. 348–52 (from the Report of the Ad Hoc Committee on Standards Activities).
61. Council, *Minutes,* September 28, 1963, pp. 336–37.
62. Council, *Minutes,* March 19, 1977, p. 96.
63. Council, *Minutes,* February 3, 1962, p. 11; *CEP,* v. 58 (March, 1962) p. 140.
64. *CEP,* v. 67 (August, 1971) p. 1; v. 74 (July, 1978) p. 96.
65. Council, *Minutes,* November 25, 1972, p. 464; see also Council, *Minutes,* June 2, 1973, pp. 225–26.
66. *CEP,* v. 74 (July, 1978) pp. 94, 96.
67. Council, *Minutes,* November 30–December 1, 1963, p. 646.
68. Council, *Minutes,* March 9–10, 1963, p. 35; November 30–December 1, 1963, pp. 646–49; Executive Committee, *Minutes,* March 24, 1961, pp. 106, 113–14.
69. Executive Committee, *Minutes,* March 24, 1961, pp. 113–14 (Appendix D).

70. Executive Committee, *Minutes*, March 24, 1961, p. 106; Council, *Minutes*, August 26, 1961, pp. 200, 214-15.
71. *CEP*, v. 60 (January, 1964) pp. 118-19.
72. Council, *Minutes*, April 22, 1960, p. 70; June 18, 1960, pp. 86-87, pp. 153-55.
73. Council, *Minutes*, September 24, 1960, p. 183.
74. Council, *Minutes*, May 19, 1962, p. 96.
75. Council, *Minutes*, August 25, 1962, p. 180.
76. Council, *Minutes*, March 9-10, 1963, p. 35.
77. Council, *Minutes*, March 9-10, 1963, pp. 35-36.
78. Council, *Minutes*, December 4, 1965, p. 463.
79. Executive Committee, *Minutes*, January 15, 1966, pp. 40-41.
80. Council, *Minutes*, March 9-10, 1963, pp. 28-29.
81. Council, *Minutes*, March 9-10, 1963, p. 29.
82. Council, *Minutes*, May 4-5, 1963, pp. 144-45.
83. Council, *Minutes*, July 18, 1964, pp. 329, 344.
84. Council, *Minutes*, August 23, 1969, pp. 277-78 (for Research Committee); June 2, 1973, p. 225 (for Machine Computation Committee).
85. Council, *Minutes*, February 20, 1960, pp. 39-44. See also the survey of the South Texas Section in *CEP*, v. 55 (July, 1959) p. 74.
86. Council, *Minutes*, January 10, 1959, p. 117; Executive Committee, Minutes, February 21, 1959, p. 121.
87. Executive Committee, *Minutes*, February 21, 1959, p. 121.
88. Executive Committee, *Minutes*, April, 18, 1959, p. 242; Council, *Minutes*, May 16, 1959, p. 269. *CEP*, v. 57 (October, 1961) p. 92, has a brief history of the committee.
89. "Dynamic Objectives for Chemical Engineering, 1961," *CEP*, v. 57 (October, 1961) pp. 69-99. As noted in the previous reference, a brief history of the Dynamic Objectives Committee is included on p. 92. For the award see *CEP*, v. 58 (June, 1962) p. 17.
90. William C. Schall, "What ARE the Dynamic Objectives," *Chemical Engineering*, v. 68 (March 20, 1961) pp. 166-70; "AIChE Report Now Gives Clear Call for Major Reform," *ibid.* (December 11, 1961) pp. 158-62.
91. Engineers' Joint Council, *The Engineering Profession in Transition*, New York, 1947, p. 14.
92. See, for example, *CEP*, v. 55 (March, 1959) p. 138; v. 57 (March, 1961) p. 105; v. 56 (September, 1960) p. 33; v. 61 (January, 1965) p. 53. Also Executive Committee, *Minutes*, November 18, 1961, pp. 381-84.
93. Executive Committee, *Minutes*, November 18, 1961, p. 381; Council, *Minutes*, September 23, 1961, p. 266; and *CEP*, v. 57 (November, 1961) p. 23, for example.
94. Council, *Minutes*, September 23, 1961, pp. 257, 266-68 (Appendix C).
95. Council, *Minutes*, December 2, 1961, p. 395.
96. Council, *Minutes*, February 3, 1962, pp. 12, 21-28.
97. Council, *Minutes*, May 19, 1962, p. 93.
98. Executive Committee, *Minutes*, February 16, 1963, p. 9.
99. Council, *Minutes*, May 16, 1964, p. 218.
100. Council, *Minutes*, May 16, 1964, p. 218; *CEP*, v. 60 (June, 1964) p. 15.
101. *CEP*, v. 62 (July, 1966) p. 218.
102. Executive Committee, *Minutes*, October 3, 1969, p. 362; April 12, 1970, p. 112; *CEP*, v. 67 (April, 1971) p. 136.
103. Marvin Fleischman, "Are You Ready to Talk About Chemical Engineering," *Chemical Engineering*, v. 85 (September 11, 1978) p. 171.
104. *CEP*, v. 60 (August, 1964) p. 164; Executive Committee, *Minutes*, June 29, 1968, p. 390.
105. *CEP*, v. 55 (November, 1959) p. 37.
106. Monroe W. Kriegel, "Economics of Continuing Education," *CEP*, v. 62 (August, 1966) p. 21, briefly reviews the history of concern about technological obsolescence.

EXPANDING PROGRAMS

107. "Dynamic Objectives for Chemical Engineering, 1961," *CEP,* v. 57 (October, 1961) p. 72.
108. W. Robert Marshall, "Continuing Education in Chemical Engineering," in W. T. Koetsier, ed., *Chemical Engineering in a Changing World,* Amsterdam, 1976, p. 450.
109. John J. McKetta, "What Can a Professional Society Do?," *CEP,* v. 59 (October, 1963) p. 23.
110. Marshall, "Continuing Education in Chemical Engineering," p. 445.
111. Council, *Minutes,* May 4-5, 1963, p. 148.
112. Executive Committee, *Minutes,* June 22, 1963, p. 255.
113. Executive Committee, *Minutes,* November 2, 1963, p. 499.
114. Executive Committee, *Minutes,* July 26, 1966, p. 220.
115. Marshall, "Continuing Education in Chemical Engineering," p. 450.
116. Council, *Minutes,* November 30-December 1, 1963, p. 557.
117. S. W. Churchill, "Once More, With Feeling," *CEP,* v. 63 (August, 1967) p. 23.
118. "Controversial ASEE Report Emerges Under Fire," *CEP,* v. 61 (November, 1965) pp. 39-40; W. B. Franklin, "ASEE: The Report and the Story Behind the Report," *CEP,* v. 62 (February, 1966) pp. 18-21. For typical reactions to the report see *CEP,*v. 61 (November, 1965) pp. 40-43; v. 62 (April, 1966) pp. 112-13; v. 62 (February, 1966) pp. 21-27; v. 63 (August, 1967) pp. 13-35.
119. *CEP,* v. 62 (March, 1966) pp. 40-41; v. 63 (August, 1967) p. 36; v. 64 (December, 1968) pp. 25-26.
120. These forums are reported in detail in "ASEE: The Sound and the Fury," *CEP,* v. 62 (February, 1966) pp. 17-49, and "ASEE Goals Report: Round Two," v. 63 (August, 1967) pp. 13-35.
121. *CEP,* v. 62 (March, 1966) p. 40; v. 63 (August, 1967) p. 36.
122. *CEP,* v. 62 (April, 1966) pp. 112-13.
123. "ASEE Goals Report: Round Two," *CEP,* v. 63 (August, 1967) pp. 13-35.
124. *Chemical Engineering,* v. 74 (June 5, 1967) p. 63; "ASEE Goals Report: Round Two," *CEP,* v. 63 (August, 1967) pp. 28-35.
125. *Chemical Engineering,* v. 74 (June 5, 1967) p. 63; *CEP,* v. 63 (August, 1967) p. 35.
126. "Goals Report Gutted," *Chemical Engineering,* v. 75 (November 4, 1968) pp. 80-82.
127. Council, *Minutes,* February 19, 1967, p. 267; November 25, 1967, pp. 607, 646-94; and Executive Committee, *Minutes,* November 11, 1967, pp. 586-87, for details of the EJC constitutional change and AIChE's reaction to it.
128. Executive Committee, *Minutes,* January 20, 1968, p. 3, and personal correspondence, Van Antwerpen to author, December 6, 1982.
129. Council, *Minutes,* November 10, 1973, p. 505.
130. For instance, William H. Wisely, "The Influence of Engineering Societies on Professionalism and Ethics," in *Ethics, Professionalism, and Maintaining Competence,* New York, 1977, p. 51, notes that codes of ethics have not enjoyed high priority in most engineering circles.
131. For the history of engineering codes of ethics see Hunter Hughes, "Engineering Ethics: The Search for a Single Code," *Consulting Engineer,* v. 15 (July, 1960) pp. 112-21, and Wisely, "Influence of Engineering Societies on Professionalism and Ethics," pp. 51-62.
132. Executive Committee, *Minutes,* June 17, 1961, p. 161.
133. Council, *Minutes,* December 5, 1964, pp. 444, 448; May 14, 1966, pp. 153-54.
134. Executive Committee, *Minutes,* August 20-21, 1965, p. 292.
135. Council, *Minutes,* September 25-26, 1965, pp. 339-40; December 4, 1965, p. 402.
136. Council, *Minutes,* December 4, 1965, p. 402; Executive Committee, *Minutes,* January 15, 1966, p. 44.
137. A questionnaire published in *Chemical Engineering* yielded a majority of more than 80% favoring engineering societies establishing a review board for ethical problems. See Roland A. Labine, "Engineers Ask for Action!," *Chemical Engineering,* v. 72 (February 15, 1965) p. 190. But a questionnaire asking for responses to individual ethics problems yielded a wide diversity of

opinion and little consensus on many issues. See "Engineers Speak Out on Ethics," *Chemical Engineering*, v. 70 (December 9, 1963) pp. 177–84.
138. For example, "Ethics in Business," *CEP*, v. 61 (February, 1965) pp. 13–25; "Recruiting Practices & Ethics: An Introduction," *CEP*, v. 64 (September, 1968) pp. 27–46; "How Confidential Should a Confidentiality Clause Be?," *CEP*, v. 66 (July, 1970) pp. 23–46. See also *CEP*, v. 76 (December, 1980) pp. 15–23; v. 77 (January, 1981) pp. 24–30; (February, 1981) pp. 22–25.
139. Robert Perrucci and Joel E. Gerstl, *Profession without Community: Engineers in American Society*, New York, 1969, p. 129.
140. Council, *Minutes*, March 15, 1969, pp. 61, 95–98; May 3, 1969, pp. 132–33, 155–57.
141. At the Council meeting of March 15, 1969, one Council member reported that he "had taken a poll of his company and the reaction [to the model confidentiality clause] was negative and that this was a sensitive area for AIChE involvement." Council, *Minutes*, March 15, 1969, p. 61; see also Council, *Minutes*, May 3, 1969, p. 132.
142. Council, *Minutes*, May 3, 1969, pp. 132–33.
143. Council, *Minutes*, August 23, 1969, pp. 343–45; November 28, 1970, pp. 400, 478–85; "Guidelines for Confidentiality Clauses," *CEP*, v. 67 (January, 1971) pp. 17–19.
144. Dimitrios Tassios, "What's Up with Pensions," *Chemical Engineering*, v. 79 (August 7, 1972) p. 86.
145. For a discussion of the issue see "Pensions & Professionalism," *CEP*, v. 62 (September, 1966) pp. 27–41.
146. *Chemical & Engineering News*, v. 43 (October 25, 1965) pp. 70–77; *CEP*, v. 62 (September, 1966) p. 33.
147. Roland A. Labine, "Engineers Ask for Action!," *Chemical Engineering*, v. 72 (February 15, 1965) p. 190.
148. Council, *Minutes*, May 15, 1965, p. 73; *CEP*, v. 62 (September, 1966) pp. 27–28; v. 64 (March, 1968) p. 41.
149. *CEP*, v. 62 (September, 1966) p. 27; v. 64 (March, 1968) p. 41.
150. *CEP*, v. 62 (September, 1966) pp. 38–39.
151. Council, *Minutes*, December 3, 1966, pp. 402–03, 473–74; Executive Committee, *Minutes*, April 14, 1967, p. 21.
152. Council, *Minutes*, September 23–24, 1967, pp. 562–64.
153. Council, *Minutes*, September 23–24, 1967, p. 537.
154. Edwin T. Layton, Jr., "Past Attempts to Unify Engineering Professionals," in *Ethics, Professionalism, and Maintaining Competence* (ASCE Conference, March 10–11, 1977), New York, 1977, pp. 132–46, points to the inability of engineers to reach consensus in non-technical matters as a major factor in the lack of unity in the engineering profession as a whole.
155. Council, *Minutes*, May 18–19, 1968, p. 227, summarizes the background of this proposal.
156. Council, *Minutes*, May 18–19, 1968, pp. 151–53, 227–36.
157. Council, *Minutes*, September 21, 1968, pp. 468, 538–48.
158. Council, *Minutes*, September 21, 1968, p. 468; *CEP*, v. 64 (December, 1968) pp. 84, 86.
159. Council, *Minutes*, March 15, 1969, pp. 69–70.
160. Council, *Minutes*, May 3, 1969, p. 125.
161. Council, *Minutes*, May 3, 1969, pp. 144–49.
162. Executive Committee, *Minutes*, June 21, 1969, p. 227.
163. Council, *Minutes*, August 23, 1969, pp. 227, 305–06.
164. Council, *Minutes*, February 14, 1970, p. 46; *CEP*, v. 72 (February, 1976) p. 28.
165. Baxter L. Tilford, "Engineering and Disadvantaged Youth," *CEP*, v. 72 (February, 1976) pp. 28–29.
166. Council, *Minutes*, February 1, 1964, p. 24.
167. Executive Committee, *Minutes*, November 14, 1964, pp. 427–35.

168. Council, *Minutes,* December 5, 1964, pp. 454–55; February 6, 1965, p. 27.
169. Council, *Minutes,* May 3, 1969, pp. 125–26.
170. T. A. Burtis, "Speaking Out on Public Issues," *CEP,* v. 63 (December 1967) p. 31.
171. Council, *Minutes,* March 10, 1973, p. 170.
172. Council, *Minutes,* February 25, 1961, pp. 48–49.
173. Executive Committee, *Minutes,* January 21, 1961, p. 5; Council, *Minutes,* February 25, 1961, p. 18.
174. Council, *Minutes,* August 25, 1962, pp. 184, 206–17.
175. For AIChE's increased international relations activities, see, for instance, *CEP,* v. 61 (August, 1965) p. 47; v. 61 (October, 1965) p. 166; v. 63 (September, 1967) p. 154; and v. 66 (April, 1970) p. 49.
176. *CEP,* v. 60 (June, 1964) p. 18.
177. *CEP,* v. 61 (August, 1965) p. 47.
178. *CEP,* v. 59 (May, 1963) pp. 24–25; v. 60 (June, 1964) p. 164.
179. Council, *Minutes,* November 25, 1972, pp. 518–21.
180. Council, *Minutes,* June 2, 1973, p. 223.
181. Executive Committee, *Minutes,* April 12, 1970.
182. Gloria M. Lambson, "Forecasting Employment Demand for Chemical Engineers," Master of Business Administration thesis, New York University, October, 1976, p. 19. The estimates of others are somewhat lower. For instance; Ralph Landau, "The Chemical Engineer—Today and Tomorrow," *CEP,* v. 68 (June, 1972) p. 9, gives figures suggesting that between 51% and 60% of all chemical engineers were members of AIChE.
183. *CEP,* v. 49 (November, 1953) p. 23, n. 7, mentions that 59% of AIChE members were also members of ACS in 1947, 51% in 1953. The estimate of one-third in the mid 1960s came from the AIChE's former Executive Secretary/Director, F. J. Van Antwerpen, January 21, 1982. According to the present Executive Director of the AIChE, J. Charles Forman, today less than 10% of ACS members are also chemical engineers.
184. "Report of the Ad Hoc Committee on the Future of the Institute," in Executive Committee, *Minutes,* April 14, 1967, p. 70.

CHAPTER 4

Maintaining Balance, c1970–c1983

Although the 1960s saw a shift of the Institute towards increased involvement in non-traditional programs, this shift was quite limited. The early 1970s, however, saw the organization become much more deeply involved in several such programs, particularly in the areas of employment conditions and government interaction. Two factors were primarily responsible for this: the severe engineering recession of 1970–72 and the rapid growth of the federal government's involvement in regulatory activities affecting the chemical industry. The former significantly increased the pressure on AIChE to take a more aggressive role in promoting the economic status of chemical engineers, while the latter provided the stimulus for increased interaction with governmental agencies.

THE RECESSION OF 1970–72 AND AIChE JOB-RELATED PROGRAMS

In the late 1960s and early 1970s the engineering profession was hit by a severe unemployment crisis. Massive federal reductions in funding for aerospace and defense projects led to heavy engineering layoffs. On the heels of these layoffs came a severe economic recession, which made matters worse. For the first time in recent history unemployment became a major concern of engineers. The Engineer/Scientist Demand Index of Deutsch, Shea & Evans indicated the severity of the problem. The index, based on the volume of advertising space directed towards engineers and scientists in newspapers and technical journals, was begun in 1961 and based on 1961 = 100. The index had risen steadily and had peaked in 1966 at about the 200 level. But it had begun to drop in the late 1960s with the decline of federal commitment to technological and scientific research and development. The recession accelerated in the fall, and by December, 1970, it was at 35.3, an all-time low.[1]

Although chemical engineers had survived the Great Depression of the 1930s relatively unscathed, the profession was significantly affected by the 1970–72 recession (although not to the same extent as fields like electrical engineering). As early as June, 1969, *CEP* published a letter reporting large layoffs by several major employers of chemical engineers.[2] By 1970 tales of major layoffs were circulating in the profession.[3] There was some truth in the tales. The "normal" unemployment rate among chemical engineers was somewhere under 1%.[4] In January, 1971, an AIChE survey found that about 1.7% of all AIChE members were unemployed, and the situation continued to deteriorate.[5] In May, 1971, AIChE Vice President Tommy Tomkowit reported that 61% of chemical engineering graduates at the B.S. level had no job offers,[6] and by August, 1971, the unemployment rate among AIChE members had risen to 3.7%, at least four times higher than normal.[7]

The recession of 1970-72 produced a crisis in many of America's major engineering and scientific societies as their members, unemployed or feeling threatened by unemployment, began to demand more job-oriented programs, even at the cost of traditional technical programs. In 1973, 87% of the respondents to an *Industrial Research* poll declared that technical societies were not providing adequate job-oriented support for their members, and 52% said they would consider joining a collective bargaining unit.[8] This discontent soon began to have an impact. In the American Chemical Society members demanding more society involvement in employment conditions elected a petition candidate, Alan C. Nixon, as president. Shortly after, the ACS under Nixon's leadership embarked on an ambitious but controversial program of establishing and attempting to enforce employment guidelines for employers of chemists.[9] Electrical engineers, hard hit by government reductions in defense and space spending, reacted similarly. A constitutional amendment which would have transformed the Institute of Electrical and Electronics Engineers from a scientific organization into an organization focused more on economic matters was narrowly defeated: 23,633 to 23,266. In response to this pressure the IEEE began to increase its activities in non-technical areas in the early 1970s, initiating a special dues assessment of $5 in 1973 to cover the costs of these programs.[10] They also voluntarily changed their tax status from an educational society to a business league. The American Society of Mechanical Engineers also modified its course in the early 1970s, adopting the goal of making ASME responsible to the economic and other non-technical needs of its members, as well as to their technical needs.[11]

Like most American engineering societies, AIChE, prior to 1970, had been only marginally involved in problems focusing on its members' job-related needs. It had published a short pamphlet on professional standards which contained some sections dealing with the employer's responsibilities to his professional employees in the mid 1950s. It had established a small employment service in the late 1950s and had begun participation in the American Chemical Society's study of portable pensions in the late 1960s. In addition, the AIChE had begun offering group life insurance to members in 1967.[12] But none of these activities absorbed a large amount of the Institute's finances or energies.

Even in relatively prosperous times, non-supervisory and lower management chemical engineers often felt that programs such as these were insufficient and that AIChE and other technical societies were not adequately serving their non-technical (i.e., economic) needs. In 1964 over 80% of those responding to a poll held this position, and 60% maintained that engineering societies should become involved in setting minimum standards of employment for engineers.[13] These feelings remained relatively dormant during the prosperous years of the 1960s but erupted during the lean years of the 1970-72 recession.

In the early 1970s, irritated by the rumors of massive layoffs where inadequate notice and respectable separation allowances had not been provided and feeling threatened themselves, a growing number of members began to demand that AIChE shift from being a strictly learned society to one that would be more active in disputes between employers and employees and would actively seek to improve the economic status of chemical engineers in areas like pensions, job security, employment termination, and working conditions.[14]

As early as April of 1970 the Executive Committee discussed the fact that a larger number of letters than usual were coming in from chemical engineers claiming that "the AIChE does nothing for the working member" and resigning their membership.[15] As the employment crisis worsened dissidents began to charge that the AIChE was dominated by university professors and corporate management who favored oversupply and unemployment.[16] University professors, they argued, needed large numbers of students to maintain their academic empires and to use for labor on research projects; industrial executives favored oversupply since it depressed engineering salaries. Critics cited the continued circulation of the career guidance film produced in the early 1960s as evidence of continued proselytization by the AIChE even in an era of unemployment.[17]

One of the most widespread demands of dissident chemical engineers was that AIChE publish a list of employers who engaged in unethical personnel practices.[18] The intensity of the feelings of some members on this issue can be judged from a letter published in *CEP*:

> This would mean enforcing discipline within the membership ranks, and prohibiting members from accepting employment with a black listed company. I realize this approach smacks of unionism, but so be it. Maybe we have been the nice guys too long.[19]

The recession clearly brought a significant shift in sentiments toward unionization among AIChE members. In 1959 only 6.2% had favored collective bargaining by a professional union.[20] In a 1973 survey, 35% held that the individual engineer needed a union, and 30% felt that unions were the only effective way to keep engineers abreast of the gains made by other employee groups.[21]

Throughout the early 1970s AIChE's leadership had to carefully balance their actions. To satisfy the growing demands of lower management and non-supervisory engineers, the Institute initiated new programs and considerably expanded old activities in job-related areas, including: (1) employment services, (2) economic surveys, (3) pension studies, and (4) employment guidelines. But Council attempted to formulate and implement these programs in a manner tolerable to its upper management members and to those engineers who continued to be more concerned with traditional technical programs.

Employment Services

The AIChE had begun to create an effective employment service for members in 1956 at the recommendation of Van Antwerpen, its Executive Secretary. It had begun by mailing a reprint of *CEP*'s employment section ("Situations Wanted") to employers to give it wider circulation. In 1957, to make this service more effective and to get it to prospective employers more quickly, it began issuing its "Situations Wanted" section as a preprint. In 1959 the Institute added James Buchanan to its staff as Employment Coordinator. That same year it issued a book of employment pointers and a list of technical employment agencies for unemployed members, and in 1961 it began preprinting *CEP* recruitment ads and forwarding them, two weeks ahead of normal publication, to members seeking employment. To provide additional help, the Institute in 1959 joined the Engineering Societies Placement Service, a non-profit employment agency sponsored by the Founder Societies.

MAINTAINING BALANCE 93

The decision to join the ESPS did not affect the growth of AIChE's employment service. The Institute decided to continue to operate it, a decision which proved fortuitous when the ESPS was liquidated in 1965, unable to compete with private employment agencies when employment conditions were rosy. In 1963 AIChE began to monitor recruitment advertising in over 80 newspapers and technical publications, using this monitoring service to build its mailing list for preprints of *CEP*'s "Situations Wanted" section. Finally, as part of the employment service, all members were permitted two free "Situations Wanted" advertisements per year.

Because it already had an employment service in place, the Institute was able to respond quickly to the unemployment crisis and to the growing demands for programs focusing on the economic needs of chemical engineers by expanding its programs.[22] In early 1971, a network of local section employment coordinators was established. These served as clearing houses for job-related information in their local areas and as a focal point for job-related information provided by the national AIChE employment service. By August, 1972, 54 local sections had appointed employment coordinators who received job leads culled by the national's employment monitoring service three days a week. The same packets were offered directly to unemployed members in regions not close to local sections for a nominal fee. The mailing list for *CEP*'s preprint on "Available Chemical Engineers" was expanded to 9,000 names. Finally, beginning in 1971, the Institute permitted unemployed members to insert an indefinite number of "Situations Wanted" notices in *Chemical Engineering Progress* and in the preprinted AIChE employment bulletins.[23]

The expanded AIChE employment service was apparently effective during the 1970–72 recession. Two pieces of evidence support this. First, *CEP* in March, 1972, published a modestly large collection of testimonials praising the service.[24] Second, in June, 1972, AIChE's Executive Committee voted to limit the availability of the service to chemical engineers who had been dues paying members for at least a year because unemployed nonmembers were joining the Institute, taking advantage of the postponed dues offered unemployed engineers (beginning in February, 1971) to gain access to AIChE employment services.[25]

Economic Surveys

In 1970 the AIChE launched a second major program in the economic area: economic benefit surveys. This program, like the employment service, was not completely new. In 1953, in conjunction with a broader survey of membership attitudes, the Institute had surveyed member earnings.[26] The AIChE's Executive Committee had considered regular salary surveys, but rejected the idea because other groups, like the Engineering Manpower Commission, were already making surveys of the earnings of engineers (though not chemical engineers specifically).[27]

Prodded by the growing discontent among membership with AIChE services in the employment area, Van Antwerpen suggested to the Executive Committee in June, 1970, that AIChE initiate surveys of its members to gather solid information on the economic status of chemical engineers.[28] Initial reservations were countered by the realization that many of the objections to the survey could be eliminated if the results were kept confidential (that is, released to members only and not publicized).[29] Council authorized a national economic survey of membership in November, 1970.[30]

The first economic survey was distributed to a statistically balanced cross section (25%) of AIChE members in January, 1971. It appraised such factors as employment, salaries, unionization, pension coverage and eligibility requirements, company size, supervisory responsibility, and function on a regional as well as a national basis. The results indicated that the unemployment rate for chemical engineers was 1.7% overall, but above 3% in several regions, particularly the West and Northeast. On the positive side, however, it indicated that 80% of the Institute's members were covered by pensions and that less than 6% were in companies with no pension plans.[31]

Membership reaction to the economic survey was highly favorable. This, and the deteriorating economic picture (an interim unemployment survey in July, 1971, indicated an unemployment rate of 3.7%),[32] led to a second economic survey in January, 1972. By this point conditions had begun to improve; the unemployment rate among chemical engineers had dropped to 2.8%.[33]

Because of membership's favorable reaction the AIChE economic survey, improved and expanded, became a regular annual offering. It proved valuable in subsequent years because it provided hard facts about the economic status of the profession and because it proved to be of continuing interest to employers and individual members, becoming one of the organization's best "selling points."[34]

The employment service and annual economic benefits surveys were successful employment-oriented programs expanded by or initiated by AIChE in response to the 1970–72 recession. The Institute's venture into the pensions area, however, did not enjoy similar success.

Pensions

In the mid 1960s, as noted in the previous chapter, membership concern about the long vesting periods required for some industrial pension plans had led Council to appoint a liaison to the American Chemical Society, which was much farther advanced in its consideration of portable pensions. By 1971, the ACS had created a profession-wide pension plan called "Pensions for Professionals" (PFP) and had invited AIChE to join and provide seed money. Elements of AIChE's leadership were disappointed in the plan since it provided neither a truly portable pension nor immediate vesting. Employees were still required to achieve vesting (five years were required for full vesting) with every employer.[35] Nonetheless, Council in February, 1971, approved "in principle" AIChE's participation in the plan, but limited the society's financial commitment to $10,000.[36] Two factors account for this decision. First, membership pressure for more active involvement in economic activities was strong due to the employment crisis. Second, the plan had potential benefits to those members not covered by any existing pension plan. The results of the first economic survey, however, were soon to indicate that only 5 to 10% of AIChE members fell into this category.[37]

Over the next few years the ACS struggled to improve and implement its plan. But it faced two severe problems—one financial, one legal. The costs of operating a portable pension plan were too high, and the Internal Revenue Service prohibited discrimination in favor of one class of employees over another (i.e., it prohibited early vesting for engineers and scientists unless similar privileges were offered all classes of employees). The legal problem was eliminated in 1974 when Congress passed legislation permitting creation of multi-employer pension plans for engineers. (A Joint Pension Committee of professional societies, including AIChE, was instrumental in securing this legislation.)[38] But the other problem—the high cost of operating a portable pension plan—proved insoluble.

Thus, even though AIChE remained true to its original commitment to the ACS plan,[39] by 1977 the Institute freely acknowledged that PFP had been "unable to achieve its object."[40]

Employment Guidelines

The AIChE's expanded employment aids program, its new economic benefit surveys, and its increased participation in programs to develop and improve pensions were regarded by many of the younger members as commendable but inadequate responses to the problems faced by engineers employed in industry. They felt that the Institute should more actively attempt to improve promotion policies, salaries, termination conditions, and working conditions. As the recession worsened in 1970 and early 1971 pressure on the organization increased. The Sacramento Valley Section, for instance, passed resolutions in early 1971 calling for Council to develop and publish a system rating employers on their attitudes toward chemical engineers.[41] The Pittsburgh Section passed resolutions calling for increased action not only on portable pensions but on employment contracts, secrecy agreements, and patent rights. The resolutions of both of these sections were endorsed "in principle but not detail" by the Cleveland Section and forwarded to national headquarters.[42]

In response to these pressures, the Institute embarked on yet another major employment-oriented program: the development and implementation of a set of employment guidelines for chemical engineers. As early as October, 1970, after approving the economic survey, the Executive Committee reviewed what else the Institute could do for its members in the professional, economic, and employment areas and considered the possibility of revising the 1955 "Professional Standards" pamphlet.[43] At the November, 1970, Council meeting, Vice President J. J. Martin argued that simply revising the old "Professional Standards" pamphlet would not be enough, that what was needed was a completely new AIChE statement on employment practices.[44] In January, 1971, Council turned this task over to its Professional Development Committee.[45]

In May, 1971, P. D. Lederman, chairman of this committee, reported to Council that he hoped to have a preliminary set of employment guidelines ready before September, 1971, when he had scheduled a meeting between his committee and around 50 representatives from top level management to receive their criticism and input. A final committee report would then be prepared, and it would be submitted to Council in November, 1971.[46]

The Professional Development Committee met its November deadline. Council received, reviewed, and accepted in principle the guidelines the committee had developed at that time.[47] At its February, 1972, meeting Council reviewed an edited version in more detail. Several directors expressed reservations about the guidelines, complaining that they did not sufficiently emphasize employee responsibility and that they were too specific in the areas of pensions and terminations. Council, nonetheless, voted unanimously to adopt the guidelines, publish them in *CEP*, put them into booklet form, and begin discussion of implementation after presentation to the membership for discussion.[48]

On the whole, AIChE's "Guidelines to Professional Employment" was a moderate document.[49] The sections on recruitment, terms of employment, professional advancement, and professional development contained few controversial items. Most of the obligations the "Guidelines" spelled out were obligations generally accepted by the profession and by industry as fair practice. For instance, under professional advancement the "Guidelines" recommended that engineers receive performance reviews from their supervisors on an annual basis and that engineers whose aptitudes and interests were technical rather than supervisory be given equivalent means of advancement and recognition.

The sections on transfer or termination and on pensions and retirement, however, reflected the concerns of the times and were far more specific. They recommended notice of one to three months for transfer and one month for termination. They suggested severance pay equal to at least one-half month's salary plus two weeks of salary per year of service. The "Guidelines" on retirement recommended that engineers become eligible for pension plans after no more than a year of service and that they become fully vested within three years.

The response of AIChE members to the "Guidelines" varied widely. In many cases it was apathetic. In August, 1972, for example, President Tomkowit reported that he had received only 15 responses from the membership on the content of the "Guidelines," 9 of them favorable.[50] The Professional Development Committee had a somewhat better response.[51] Membership attitude toward the document is probably best judged by the opinion poll carried out by the society in 1975. The poll indicated that around 52% of the organization's members had at least some familiarity with the "Guidelines" content, although more than half of these were "not too familiar." But 76% felt, after reading them, that they were "about right" in terms of content and philosophy, and almost 90% felt that every company should establish policies consistent with them.[52]

Since the "Guidelines" represented the insertion of a new party in employer–employee affairs, industry's response was very guarded. In August, 1972, Tomkowit noted that he had sent copies to 250 employers. Of the 44 that replied, 24 were generally favorable, 16 were relatively neutral, 4 were unfavorable. Nine companies, however, "very specifically" objected to the sections dealing with severance pay and vesting, feeling that they were too specific and imposed too many financial burdens on employers.[53]

After accepting and ordering publication of the "Guidelines" in February, 1972, Council began to study implementation. The AIChE, first, solicited input from members and held an open forum on the subject at its May, 1972, meeting in St. Louis.[54]

Joint Guidelines

. At the same time, the Institute was cooperating with other professional societies attempting to produce a common set of employment guidelines since several other societies had previously prepared their own guidelines or were in the process of preparing them when AIChE began its efforts. In October, 1971, after earlier contacts by AIChE President J. J. Martin, representatives from AIChE met with representatives from ASME, ASCE, ACS, IEEE, and the National Society of Professional Engineers (NSPE), and asked them to cooperate in developing a set of mutual guidelines.[55] The efforts of most of these groups were eventually channeled through a joint committee of the Engineers' Joint Council and NSPE, co-chaired by two AIChE members: Irving Leibson, AIChE Vice President, representing EJC; and E. E. Slowter, a member of both AIChE and NSPE, representing NSPE.[56] The guidelines this committee produced, called the "Joint Guidelines," were approved by AIChE's Executive Committee in February, 1973.[57]

These did not differ in most areas from AIChE's original "Guidelines." But the sensitive sections on severance pay and pensions were toned down. For example, AIChE's "Guidelines" had recommended full vesting in pension plans within three years; the joint document recommended full vesting in five years. Recommended termination notice and severance pay were also significantly reduced.[58]

While the "Joint Guidelines" were being prepared, the AIChE continued to debate the problem of implementation. Opinions varied widely. A vocal group of members wished to pressure employers into compliance by publishing lists of those who had promised to subscribe to the "Guidelines," those who had refused to subscribe, and those who had promised to subscribe but had violated their promises.[59] Industry's attitude towards implementation was at the opposite end of the spectrum and was clearly expressed by Eugene Barr, Executive Vice President of Union Carbide. Industry, he declared, would give "serious consideration" to the "Guidelines" when establishing or revising personnel policy, but it would not accept a system where a third party (like AIChE) passed judgment on a company's compliance.[60]

The implementation plan submitted to Council by the Member Relations Committee and adopted in August, 1972, tried to steer a middle path. The committee rejected the use of threats or public embarrassment to coerce employers into compliance, arguing that tactics such as black lists would simply push employers into defensive verbalizing and that judgment of compliance was too troublesome, since acceptance of the "Guidelines" in every small detail was nearly impossible and acceptance in principle would be difficult to measure. The AIChE, the committee concluded, could best secure compliance with the "Guidelines" through an aggressive, continuing program to make both chemical engineers and their employers aware of its principles and through provision of a forum for publicizing employment practices that gave engineers a greater sense of professionalism.[61]

The AIChE actively sought to implement this program over the next five years. By the fall of 1972 the "Guidelines" had been sent to 290 major companies in the chemical processing industry, and when the "Joint Guidelines" were adopted the Institute mailed 10,000 copies to industrial executives all over the country.[62] *CEP* published articles dealing with the "Guidelines" several times a year.[63] In 1974 the AIChE furnished all chemical engineering students with copies of the "Joint Guidelines," and, beginning in 1975, published them regularly in its *Student Member Bulletin.*[64]

The AIChE's executive luncheon program was probably, however, the most effective element of the organization's implementation program. For a number of years AIChE officers had held luncheons with local industrial executives in conjunction with national meetings to discuss matters of common concern. Beginning in 1972 this luncheon program was expanded, using the "Guidelines" as the principle topic of discussion. In 1973, for instance, Ted Weaver, then president of the AIChE, held around a half dozen executive meetings, urging adoption of the "Guidelines" to more than 100 industrial executives. The Institute's officers also attempted to sell the document to leading trade associations in the chemical industry.[65] In 1974 and 1975, for example, Institute officials met with representatives of the Manufacturing Chemists' Association, the American Petroleum Institute, and the Technical Association of the Pulp and Paper Industry and had some success in securing informal compliance agreements.[66]

The AIChE's guideline implementation program was significantly more aggressive than that of any of the major American engineering societies, but it still depended on persuasion rather than direct pressure. The American Chemical Society had taken a more militant position, rejecting the "Joint Guidelines" as too weak, publishing lists of companies not in compliance with ACS guidelines, and establishing a program to mediate disputes between employers and employees. Some segments of the Institute's membership clearly preferred this approach.[67]

The most vocal and persistent opposition to the national organization's moderate approach came from the North Jersey and New Jersey local sections. The principal spokesman for North Jersey AIChE members wanting more aggressive implementation was Dimitrios Tassios of the Newark College of Engineering. In opposing the AIChE's leadership on the question of implementation, he and the North and New Jersey sections pursued several courses of action. First, they (and Tassios in particular) argued their position at national meetings and in various publications. Second, the North Jersey Section began, on its own, to collect information on the practices of New Jersey companies in the pension area for comparison to the "Guidelines." And, third, Tassios sought election to national offices in AIChE, hoping to change national policies by democratic means. He maintained that AIChE should publish a directory of companies which would compare company policies with the "Guidelines" and provide a past history of company layoffs.[68] The Joint Professional Development Committee of the New Jersey and North Jersey sections argued in articles published in *CEP* criticizing the Institute's stand on implementation that economic issues were not minor to a professional society like AIChE, but critical and central.[69]

Because the national organization refused to publish data on specific companies for comparison to the "Guidelines," the North Jersey Professional Development Committee went further. In 1973 it collected comparative information (in the public domain) on the pension plans of the nine major employers of chemical engineers in New Jersey and began to prepare the data for publication. In August, 1973, however, AIChE's Executive Committee informed the North Jersey Section that it should stay within Council policy and not publish the data.[70]

Since persuasion had not altered the national leadership's views on implementing guidelines and since independent action by local sections was prohibited, Tassios and his supporters attempted to change AIChE policy on guideline implementation by election. They had some hope of success. A 1973 survey of AIChE members indicated that 61% felt that a directory rating companies on issues like past layoffs, salaries, and benefits would be "helpful."[71]

Three times in the 1970s he utilized the relatively lenient nomination-by-petition provisions of the AIChE constitution to run for national office. In 1973 he was nominated by petition for one of the four open Directorships and ran on a platform favoring more aggressive guideline implementation. But he was not elected.[72] In 1974 he was nominated by petition for Vice President (since 1930 de-facto President-elect).[73] Traditionally, AIChE's Nominating Committee had selected two candidates for Vice President from among members who had previously served as Directors of the Institute, feeling that a term as Director was necessary to understand the internal operation and financial (and other) limitations of the AIChE. Tassios' nomination for Vice President short-circuited this process. This was critical since the Nominating Committee's candidates, William Corcoran and Klaus Timmerhaus, both favored the moderate "advocate not antagonist" mode of implementing the "Guidelines." To provide a clearer picture as to the feeling of the majority of the members, William Corcoran withdrew from the race, permitting a face-to-face confrontation between Timmerhaus, who favored the approach adopted by

Council, and Tassios, who opposed it. Timmerhaus was elected 8104 votes (63%) to 4744 (37%).[74] This election led to a revision of the AIChE constitution, passed by a 77% majority (75% is necessary for adoption), requiring candidates for Vice President, Secretary, and Treasurer of AIChE, when nominated by petition, to have first served a term as Director.[75] In 1979 Tassios was again nominated by petition for Director and narrowly defeated.[76]

Council's determination to hold to its moderate position was undoubtedly solidified by the election results. It gained further support from several opinion polls conducted in the mid 1970s. For instance, a 1973 survey indicated that only 25% of AIChE members supported a proposal to raise dues by $5 to support a program of legal assistance for members in the area of employment contracts.[77] In 1975 another poll indicated that members were badly split on the questions of implementation. Only around 50% favored the more active and aggressive steps. Even more significantly, only 40% were willing to support a $5 increase in annual dues to support increased implementation of the "Guidelines," only 11% were willing to work on a local committee and only 6% on a national committee to help implement them.[78] Feeling that there was no clear mandate for abandoning its moderate position on implementation, Council in January, 1976, reaffirmed its policy of strongly advocating the "Guidelines," but avoiding actions that might "embarrass" specific companies.[79]

In November, 1978, Council authorized creation of a standing Committee on Professional Employment to continue implementation of the "Guidelines," previously the responsibility of *ad hoc* committees.[80] But the new committee had difficulty in attracting interest in large part because, by the late 1970s, the employment problems that had plagued chemical engineers in the early 1970s had vanished. Salaries had risen sharply; unemployment among chemical engineers had all but completely vanished. As economic conditions improved the concern of chemical engineers with employment conditions declined. For instance, the proportion of AIChE members who believed an engineering union desirable dropped from 27.1% in 1972 to 12.0% in 1981.[81]

In 1979 J. Y. Oldshue, then President of the Institute, asked Council members to annually review the activities of the committees for which they served as liaison to determine if any needed to be disbanded or restructured. These "sunset" reviews were continued as a regular activity. In early 1982, after one such review, the Professional Employment Committee was disbanded by Council because it had demonstrated insufficient activity to justify continuance. The activities for which it was responsible, notably monitoring the employment guidelines, were transferred to the Professional Development Committee.[82] The committee and the Institute were poised to respond to the challenges of the 82–83 recession.

The decision of AIChE's national leadership to prepare guidelines and use its prestige and position to aggressively advocate their adoption by industry but not to attempt to coerce or embarrass employers into adopting them was critical. It enabled the Institute to maintain a balance between the diverse interests of its membership without significantly alienating any major interest group. Failure to take any action at all in the employment crisis of the early 1970s would have alienated younger engineers in lower management and non-supervisory positions. Militant implementation of the "Guidelines" through publication of white lists, black lists, or directories or through creation of a committee to intervene in employer-employee relations would have alienated industry and upper management engineers. The loss of any of these groups would have severely crippled the AIChE and its programs. The moderate course of action adopted by the Institute enabled it to weather the crisis.

THE ENVIRONMENT, GOVERNMENT, AND THE INSTITUTE

At the same time the recession of 1970-72 was drawing the AIChE into programs focusing on the employment conditions of chemical engineers, the growth of public concern about the environment and subsequent federal regulation of the chemical industry were providing the stimuli for AIChE consideration of new or expanded programs in other non-traditional areas.

The AIChE and Environmental Concern

In the 1960s emerging fears about the impact of chemicals on the environment caused a sharp change in public attitude towards the chemical industry. The industry declined sharply in public esteem. For instance, the proportion of respondents having a favorable attitude towards the chemical industry in Opinion Research Corp. surveys declined from 55% in 1965 to 32% in 1971. In the latter year only the tobacco industry ranked lower in public esteem of the 20 industries surveyed.[83] The chemical industry became the primary target of a host of environmental critics like Ralph Nader, who asked chemical engineers to adopt a new ethic which would place responsibility to society above responsibility to employer.[84]

The AIChE had, prior to the 1960s, a sporadic history of concern with pollution. In 1921 it had formed an Atmospheric Pollution Committee, but it quickly died. Interest revived in the 1940s, and the Institute created an Industrial Waste Disposal Committee, whose name was changed in 1955 to the Pollution Control Committee.[85] This committee was modestly active in programming.

The Institute's reaction to the growth of more intense public concern over environmental issues and to demands that chemical engineers be more socially responsible varied. Some of its responses were traditional. It expanded or adjusted existing mechanisms (e.g., committees and divisions) to encourage stronger technical programs on pollution-related problems. But it also seriously considered initiating several new programs.

The growing interest in pollution problems in the 1960s clearly led to an expansion of AIChE's programming structure in the area. In 1963 the Institute created two new technical committees—the Air Committee and the Water Committee—to replace the old Pollution Control Committee, and in 1969 it added a Solid Wastes Committee to the group to insure even more comprehensive coverage.[86]

These three committees were very active. Between 1966 and 1970, for instance, 478 pollution-oriented papers were presented at 95 sessions at AIChE meetings, an average of almost 20 sessions and 100 papers per year and a massive increase over the previous five years. During the same period AIChE published at least a half dozen volumes focusing on pollution concerns in the *Symposium Series*. In addition, the Continuing Education Department had developed four pollution courses and had three more in the mill by mid 1971.[87]

In 1969-70 an *ad hoc* committee, chaired by Weaver (President, 1972), organized an Environmental Division, which absorbed the Air, Water, and Solid Wastes committees.[88]

After its formation in 1970 the Environmental Division quickly became one of the largest and most active of AIChE's divisions. Within ten months after formation it was second in size only to the Petroleum & Petrochemicals Division, and within two years had become larger.[89] The new division was also very active in programming. At the May, 1971, national meeting, for instance, 20 of 40 sessions were environment-oriented, and in 1975 it presented 52 sessions at the three meetings of the AIChE.[90] In addition, the division quickly established its own award and its own newsletter, and it made heavy use of the Institute's *Symposium Series* as an outlet for collections of papers on environmental subjects.[91]

In addition to expanding its programming and publications in the environmental area and forming an autonomous division to serve as the focus of environmental activities, the Institute made several other adjustments to its internal structure to meet the charge that engineers were not meeting their social responsibilities. In 1969 Council proposed adding a third objective to the Institute's constitution: "to serve society, particularly where chemical engineers can contribute to the public interest."[92] This constitutional change was adopted by the membership in 1970. The Institute's awards system was similarly modified. In 1973 the AIChE created an award for service to society.

The environmental concerns of the late 1960s and the 1970s also compelled the Institute to consider one other alteration in its internal structure: a modification of its Code of Ethics. In August, 1971, Council asked an *ad hoc* committee from the Environmental Division to determine whether the code needed to be modified to include the ethical responsibility of the chemical engineer for the environment.[93] The committee recommended that the AIChE issue a statement amplifying the Code of Ethics and more clearly spelling out what was required of chemical engineers in the environmental area.[94]

In February, 1972, Council debated this recommendation and related issues such as taking disciplinary action against members failing to meet environmental standards. Favorable action on these issues would have compelled the AIChE to set up a review board to judge the merits of environmental ethics cases, a move potentially fraught with controversy. After considerable debate Council tabled the matter on an eight to seven vote, reaffirming its traditional policy of avoiding detailed, specific involvement in ethical matters.[95]

The AIChE continued this policy through the 1970s and on into the 1980s. In 1974, for instance, Council again rejected, as it had in 1965–66, the idea of an ethics advisory panel.[96] And in 1980 Council adopted a modified version of the 1977 ECPD "Fundamental Principles and Fundamental Canons" as the AIChE Code of Ethics, but did not adopt a parallel recommendation that a task force be formed to develop specific interpretive guidelines for the code.[97]

Although these decisions naturally did not please some members and non-members, they were based on a realistic appraisal of the nature of a voluntary, professional engineering society. Since the American Institute of Chemical Engineers was a voluntary body whose overall effectiveness depended on balancing the interests of a number of major interest groups while avoiding the alienation of any, it risked division and membership losses if it took overly aggressive actions in areas where general consensus was not possible. The Institute's leaders chose the balanced role of providing a forum where different groups representing different interests could discuss ethical and environmental matters. The pages of *Chemical Engineering Progress* were one such forum. While *CEP* editorials were generally sympathetic with the plight of the chemical industry and critical of many federal regulatory activities, *CEP* also published letters critical of the chemical industry.[98]

Because environmental problems are often local in character, it is not surprising that attempts to expand the Institute's involvement in the environmental arena beyond traditional areas and programs emerged at the local section level. The most ambitious of these attempts originated in October, 1967, when John McKetta, Dean of the University of Texas College of Engineering, addressed the South Texas Section of the AIChE on "Your Greatest Failure." He argued that the chemical engineering profession had lost initiative and leadership in the pollution field, leaving it in the hands of a federal government not particularly sympathetic with the chemical industry. He warned that the engineering profession generally, and chemical engineers in particular, were facing perhaps their last chance to assume leadership in environmental issues, and he urged them to act quickly.[99]

In response to this challenge, A. Roy Price conducted a survey to determine what the South Texas Section should be doing about pollution. In response to his findings, in March, 1968, the section voted to form a "Pollution Solution Group," an *action* committee devoted to applying existing technology to help reduce pollution in the south Texas area.

The South Texas Pollution Solution Group immediately attracted 40 recruits and quickly initiated a number of projects. The activity which gained the most attention both locally and nationally was its civic action project: an attempt to improve the quality of Buffalo Bayou, a turbid stream flowing through downtown Houston. Another South Texas Pollution Solution action group attempted to establish contact with state and local pollution control officials and with legislators to provide a means for chemical engineering input into legislative and regulatory processes. A third group focused on education, attempting to keep the section's membership informed about pollution control techniques and both existing and impending laws and regulations. A fourth group handled public relations.[100]

The movement spread. Within three years there were 11 active AIChE local section pollution solution groups and 10 additional local sections were attempting to organize such groups.[101] This movement was primarily centered in the local sections, but the national organization provided aid and encouragement. *CEP* publicized the activities of Pollution Solution Groups.[102] The Institute furnished programming space at national meetings so that the groups could exchange ideas.[103] The Environmental Division created a national Pollution Solution Committee to provide national coordination.[104] In addition, when the movement began to lose momentum in the early 1970s, the Institute provided direct financial assistance. In February, 1971, Council agreed to cover the costs of a one-year pilot program using a retired chemical engineer as a full-time field specialist to revitalize the program.[105] Larry Cecil, long active in environmental affairs, took the position and had some success. By 1972 there were 36 Pollution Solution Groups in operation.[106]

In the 1970s, however, the focus of the groups steadily shifted. At first they were intended to be *action* groups rather than *study* groups, and civic projects like the Buffalo Bayou project attracted the greatest amount of attention. But it gradually became clear that even narrowly restricted environmental projects were beyond the financial and manpower resources of small, volunteer groups. While dedicated volunteers could study a problem, analyze it, and suggest and test a solution, they simply did not have the resources to implement their recommendations. Many of the Pollution Solution Groups became discouraged with the paperwork and delays entailed in seeking government financial support to implement their ideas and began to flounder.

As a result, many of the groups gradually shifted away from civic action projects towards governmental interaction. In this area the possibilities of effective action were somewhat greater, for local volunteer groups could provide local, state, or federal agencies with effective technical advice. By 1974 this shift was well underway. The primary activity of most of the Institute's pollution solution groups in that year was the preparation of comments on the Environmental Protection Agency's proposed effluent limitations guidelines.[107] With the shift from civic action to governmental interaction much of the enthusiasm which had inspired the early groups slowly evaporated. Thus, in 1975 the AIChE's Environmental Division had to undertake a revitalization program for its Pollution Solution Groups (soon to be called Critical Issues Committees, since AIChE's involvement with government had by then expanded beyond the pollution area).[108]

The AIChE and Governmental Interaction

Prior to 1970 governmental interaction, like employment relations, had been an area of minimal AIChE activity. Several factors account for this. First, the federal government had largely permitted the chemical industry to develop without regulation in the first half of the twentieth century and had rarely been involved in activities which daily affected the work of chemical engineers. There was, therefore, no pressing reason for AIChE to interact with governmental agencies on any continuing basis. Two other factors contributed to AIChE's avoidance of public policy issues: the engineer's traditional apathy towards such issues and the potentially divisive nature of governmental programs.

A reassessment of AIChE's tradition of non-involvement with the federal government was stimulated by the massive growth of federal regulation of the chemical industry in the 1960s and 1970s, regulation that was a political response to the growth of public environmental fears in the period. In the late 1960s and early 1970s the federal government imposed stringent controls over both liquid and gaseous effluents from chemical plants and moved into the field of occupational safety as well. Regulation was made even more comprehensive by the passage in 1976 of the Toxic Substances Control Act, which laid down detailed rules for the production, transportation, and marketing of virtually all chemicals. The agencies needed to enforce the mass of environmental, occupational, and related legislation affecting the chemical industry had, by the 1970s, become immense bureaucracies. The largest of these agencies, the Environmental Protection Agency (EPA), for example, had not even existed in 1960, yet by 1978 it had over 10,000 employees. Moreover, its decisions and regulations had a daily impact on the work of chemical engineers.[109]

As federal regulation of the chemical industry grew and as the importance of having some interaction with federal agencies like the EPA increased in the early 1970s, several units of the AIChE attempted to initiate contacts. They encountered difficulties. Hal Kaufman, Chairman of the Standards Committee of the Food and Bioengineering Division, reported in August, 1973, that for five years his committee had been trying to develop some useful interaction with the government and particularly with the Food and Drug Administration. But, he noted, "We've been frustrated in our efforts to serve the Government. We haven't been able to obtain any continuity."[110] The AIChE's Government Programs Advisory Committee (GPAC) was having similar difficulties. In March, 1973, Chairman Jack Dart argued that a committee of volunteers like the GPAC did not have the time or resources to achieve even a modestly effective program of AIChE involvement with the federal government.[111] An *ad hoc* committee which reviewed the

Institute's interaction with the federal government concurred:

> A committee of volunteers working in their spare time, with many of its members living and working outside of Washington, cannot provide the continuing contacts needed to develop effective relationships and cannot stay on top of the ever-changing Washington scene.[112]

Attempts by other Institute committees to respond quickly and effectively had been similarly ineffective. In 1971, for instance, Council had promised the Office of Science and Technology a review of techniques for the motivation of pollution control and had appointed a committee to carry out this charge.[113] The committee's report was submitted to Council only in early 1975.[114]

Because of the obvious problems that groups of volunteers had in dealing with the intricacies of interacting with the federal government and because of the growing conviction that increased federal activities in regulating the chemical industry made such an interaction vital, Dart recommended to Council in March, 1973, that it appoint an *ad hoc* committee to review the possibility of establishing a paid, professional AIChE Washington representative. Council accepted the recommendation.[115]

The resulting committee, chaired by Past President Weaver, reviewed the activities of other engineering and scientific societies and recommended in November, 1974, that the AIChE establish a full-time Washington representative who would arrange meetings between AIChE experts and government officials, monitor legislative and regulatory agency activities, coordinate AIChE's government activities with those of other organizations, identify issues and opportunities for effective AIChE input, and keep members informed of governmental activities. The committee also recommended that a new standing committee, with broader responsibilities, replace the Institute's largely ineffective Government Programs Advisory Committee. Weaver's committee suggested that Council publish its report in *CEP* and poll membership on their willingness to support a full-time representative in Washington with a dues increase.[116]

Council, on November 30, 1974, voted to implement the *ad hoc* committee's program contingent on members' willingness to support the program with an increase of $3 per year in dues and authorized a poll to determine if sufficient support existed.[117] The poll, conducted in early 1975, attracted considerable member interest. Over 50% of qualified voters cast ballots, with 8161 (47.9%) favoring a $3 increase in dues to establish a Washington office, 8887 (52.1%) opposed.[118]

The closeness of the vote indicated that this issue, like ethics and employment guidelines, had two valid sides. A Council decision to raise dues by $3 and hire a full-time Washington representative could have a negative impact on the organization, especially since a dues increase approved in 1973 had just raised dues by more than 30%. On the other hand, some members felt so strongly that a Washington presence was vital to the health of the profession that they suggested diverting money from AIChE's publications program to finance a government interaction program.[119] As with employment guidelines, Council was faced with finding a compromise to maintain a balance between the Institute's various interest groups.

In May, 1975, the AIChE's Executive Committee asked the *ad hoc* committee on government programs to analyze the problem of government interaction again and propose a new plan.[120] The committee in subsequent months contacted several groups of AIChE members to determine why they had voted the way they did, finding that many who had voted "no" did not oppose increased AIChE involvement with government, but objected to the dues increase. The committee therefore recommended that the AIChE initiate a governmental interaction program, but finance it out of existing income and thus operate it on a more modest scale than originally anticipated. Instead of hiring a full-time professional and establishing a separate office in Washington, the committee recommended hiring a part-time professional who already had an office of his own. This person would be the AIChE "presence" in Washington, make contacts in Congress and the regulatory agencies, contact the chairman of a newly organized Government Programs Steering Committee (GPSC) when a possibility for input presented itself, and coordinate AIChE activities with those of other major engineering societies. Council reviewed this proposal at its September, 1975, meeting and, with two dissenting votes, approved it for a two year trial.[121]

At its January, 1976, meeting Council reviewed the objectives of the proposed Washington activities and guidelines for governing these activities. After considerable debate, Council adopted a formal set of objectives, a set of guidelines, and an implementation plan.[122] The guidelines were prudent. AIChE's Washington representative (T. J. Hamilton, a trained chemical engineer and an AIChE member, of Wilson E. Hamilton & Associates was hired for the position in early 1976) was to identify only areas where Institute action would be effective (even if not decisive) and where the expertise of chemical engineers was important. Both AIChE's Washington representative and others speaking for the Institute were asked to limit input, as far as possible, to credible, apolitical, factual statements and to label opinion as opinion when it was offered. All representatives of the Institute were required to clearly identify the unit of the organization which was providing input (division, committee, task force, local section, individual). These restrictions were intended to give AIChE credibility in political circles and to allow the Institute to offer itself as an independent party and not simply a "lap dog" of industry.[123]

The governmental interaction program initiated in the mid 1970s, like the employment guidelines program, maintained balance between the organization's various interest groups. By adopting a more limited program than initially contemplated and financing it out of existing reserves, the AIChE's leadership avoided alienating the substantial body of membership opposed to further dues increases, while satisfying those who felt that it was urgent to have some governmental input. By largely limiting input to areas where there was widespread consensus and to the provision of factual technical expertise, the Institute minimized the divisiveness of political issues. In terms of maintaining internal cohesion, therefore, initiation of a significant program of political action was successful.

On an external basis, it is much more difficult to assess the success of its involvement in political affairs. The Institute was just one of many groups providing or attempting to provide input to or influence the decisions of the federal government. The organization's leadership, however, was largely satisfied with the results of the investment. A. S. West (President, 1977) pointed to the decision of the House Ways and Means Committee not to recommend a tax on petroleum-based and natural gas feedstocks as a specific instance where the society's input had had an important, even if not an exclusive, impact on a decision.[124] The Institute's Washington representative, Hamilton, was quite effective. In his first two years he had, for instance, arranged for AIChE's "Water Task Force" to present testimony to the House of Representatives on corrections needed in Public Law 92-500. He had secured widespread distribution in Congress of West's position paper on petrochemical feedstocks. He had arranged for publication of the Institute's statement on energy policy in the *Congressional Record*. And he had provided means for AIChE groups to provide input to a number of administrative and regulatory agencies. He had, generally, presented more ideas, identified more issues, and made more potential contacts than the Government Programs Steering Committee (GPSC) could possibly handle.[125]

In reviewing the Washington activities in February, 1978, a GPSC subcommittee noted that any shortcomings in the Institute's program were not the result of Hamilton's performance, but of implementation problems. Once again, the voluntary nature of the AIChE limited effectiveness. The GPSC, "made up of volunteers, suffers from precisely the same situations that all volunteer committees encounter," the subcommittee reported. Among its problems were long response time, slow development of appropriate actions, and difficulty in securing continuity of action. Nonetheless, the GPSC felt that AIChE should continue its Washington activities. To improve on past performance the subcommittee recommended that the Institute retain the services of Wilson E. Hamilton & Associates on the same part-time basis as previously, but increase Institute staff support for the Washington program by establishing a special consultant to develop task forces, coordinate and expedite task assignments, identify experts willing to make inputs in Washington, and review, edit, or prepare statements of technical input to governmental agencies.[126] Council approved these recommendations in February, 1978,[127] and appointed F. J. Van Antwerpen, when he retired as the Institute's Executive Director on July 1, 1978, to assume the post of part-time special consultant on governmental affairs.

Van Antwerpen's successor as Executive Director was J. Charles Forman, a native of Chicago and a chemical engineering graduate of MIT (S.B.) and Northwestern University (M.S. and Ph.D.). Before becoming Executive Director, Forman had worked for Abbott Laboratories in a variety of posts ranging from Senior Chemical Engineer in catalysis and high pressure work to section and project management positions in fermentation and antibiotic production (including project management overseas). He was serving as Director of Manufacturing Operations for the Agricultural and Veterinary Products Division when he left Abbott for AIChE.[128]

Under Forman's administration, the Institute continued to be satisfied with the returns from its modest Washington investment. In 1979, for instance, he noted that the AIChE had had "considerable success at our level of effort."[129] In 1980 AIChE began phasing out Van Antwerpen's services as Hamilton & Associates expanded its staff by adding an engineer specifically to provide for Institute needs.[130] In 1981, feeling that the services provided by the Washington office had been "excellent" and "the most cost-effective way" of providing input to governmental agencies, Council renewed its contract.[131] Martin Siegel, who had joined Hamilton & Associates' staff, became the AIChE's new Washington representative.[132]

In addition to quietly providing input to legislative and regulatory bodies at the federal level through its Washington office and at the state and local levels through contacts made by its local sections, the AIChE also made more frequent use of public position statements on political issues in the 1970s. Here, too, Council's policy was prudent. Council issued position statements only in its own name and did not claim to speak for the diverse membership of the entire organization. Council took a public position on issues only when it felt that its statements might have some effect, when the issue was of critical importance, and when the issue was one on which chemical engineers had some expertise. Finally, Council took a position only when it felt there was considerable support for such a stand among members.

MAINTAINING BALANCE 107

The most controversial of AIChE's political stands was on nuclear power. In the mid 1970s several anti-nuclear groups in California collected sufficient signatures to place an initiative on the June, 1976, California primary ballot which, if passed, would have frozen the construction of nuclear plants in the state and have compelled existing plants to gradually close down. In June, 1975, at the request of W. K. Davis, a former Director and future President of the Institute, Council first considered the issue,[133] and in September, 1975, Council appointed an *ad hoc* committee to prepare a position statement on nuclear power.[134] This statement, passed in November, 1975, declared that the national Council of the AIChE "strongly" believed that the continued use and expansion of nuclear power was essential to a healthy economy and thus opposed actions such as the California Nuclear Initiative and urged AIChE divisions and sections to oppose such actions as well.[135] In December AIChE's President and Vice President (K. E. Coulter and Klaus Timmerhaus) wrote all California members of the Institute urging them to become involved in the debate on nuclear power,[136] and in January, 1972, Council's position statement was published in *CEP*.[137] The Institute's actions were unprecedented. This was, according to one writer, "the first time that AIChE had thrown its weight behind one side in a local political dispute."[138] Further, when similar anti-nuclear measures were placed on the November, 1976, election ballots in six other states, AIChE members in those states were urged by a personal letter from Institute President Klaus Timmerhaus to vote against them.[139]

Even though all of the initiatives were defeated at the polls and even though Council's position on nuclear energy was probably supported by a substantial majority of AIChE members, there was sufficient negative reaction to indicate that involvement in governmental affairs could hurt a voluntary organization like AIChE. One 30-year member threatened to tear his membership card to bits because of the Institute's pro-nuclear stance, and a number of other members expressed dismay.[140]

There were other hazards, besides loss of membership, to heavy Institute involvement in non-technical activities. In 1977 the Manhattan office of the Internal Revenue Service began a systematic investigation of the non-profit organizations in the United Engineering Center and announced its intention to reclassify AIChE (among others) from a tax-exempt 501(c)(3) organization (educational and scientific society) to a 501(c)(6) organization (business league) on the basis of its increased activities outside of the traditional areas of publications and meetings.[141] This reclassification would have seriously damaged the Institute's prestige by placing it in the same classification with trade groups and would have increased the cost of Institute operations.

Because the vast preponderance of AIChE activities and budget allocations remained in the education and public service areas, AIChE appealed the reclassification. After losing an appeal to the regional IRS commissioner,[142] the Institute appealed to the national office. The AIChE's new Executive Director, Forman, presented the Institute's case in a convincing manner at a hearing held in Washington on November 30, 1978, for on June 24, 1979, the IRS informed AIChE that its 501(c)(3) classification had been upheld.[143]

In 1970 the American Institute of Chemical Engineers had been almost entirely a technical society. Nearly all of its resources and efforts had been focused on the development and diffusion of technical knowledge, despite sporadic efforts to develop other programs in the 1950s and 1960s. By the AIChE's 75th anniversary in 1983 this had changed. Under the stimulus of the recession of 1970-72, public concern over the environment, and the increasing power of the federal government, the Institute had developed a number of other programs on a significant scale. Between 1970 and 1980 the AIChE had carried out an intensive effort to develop and sell employment guidelines; it had made economic surveys of its membership a regular part of its program; it had established the most comprehensive continuing education program of any professional engineering society; and it had initiated an apparently permanent and successful effort to provide input to the federal government through a Washington office and to state and local governments through local section governmental interaction committees. Further, the Institute had succeeded in carrying out these programs despite being hemmed in by the diverse and voluntary nature of its membership, by the absence of widespread consensus on many issues, and by obvious financial limitations. Moreover, the Institute had initiated these activities while improving its internal structure and the level and quality of its traditional programs.

INTERNAL STRUCTURE

Both the engineering employment recession of 1970-72 and public and governmental concern about environmental matters had an affect on AIChE's internal structure, as well as on the growth of non-technical programs. Public concern about environmental issues, for instance, provided the stimulus for making service to society one of the constitutional objectives of the Institute, for the creation of a special AIChE award for service to society, and for the creation of the Environmental Division.

The employment recession of 1970-72, however, had a sharper impact on AIChE's governing structure and should be reviewed in more detail. In the early 1970s, in the midst of the recession, AIChE's Council faced charges that it did not respond quickly enough or forcefully enough to members' economic needs and that its attitudes toward employment and other problems differed from the attitudes of the rank-and-file.[144] Council responded to these criticisms in four ways:

- It attempted to make long-range planning a regular Institute activity.
- It created a committee to consider major modifications to the Institute's internal organization.
- It modified requirements for AIChE membership grades to make the Institute more attractive to younger engineers.
- It modified AIChE election procedures.

Long-Range Planning

One of Council's first reactions to criticism of its handling of the employment crisis was the appointment in 1971 of an *ad hoc* committee to develop a long-range planning system so that the Institute could be better prepared for future crises.[145] This committee, chaired by W. B. Franklin and Waldo Leggett, recommended to Council in May, 1972, that long-range planning become a regular, annual function of the AIChE, under the supervision of the Vice President. The idea was approved and given a trial run under Franklin, who prepared an elaborate system of forecasts and plans.[146] Weaver, 1972 Vice President, attempted in a more simplified manner to predict the industrial, economic, political, and educational trends that would affect the profession for the period 1973-75 and, on the basis of this forecast, set objectives for the Institute.[147] In 1973 Leibson took a different tack. He used a Delphi analysis, a technique utilizing successive questionnaires completed on a confidential basis by experts, with the combined inputs from each round of the questionnaire made available to each of the experts during successive rounds. His analysis, however, focused on the chemical engineering profession and the chemical industry, not specifically the AIChE.[148]

As the employment crisis of the early 1970s eased, the Institute's attempt to make long-range planning a regular annual function lapsed. Planning, however, continued to be conducted on an occasional basis. In October, 1974, for example, AIChE's Executive Committee began to consider the possibility of a dynamic objectives study similar to that carried out in 1961.[149]

This study, launched in mid 1975 and completed in early 1977, called for the Institute to pursue ten objectives.[150] A number of these were extensions of programs the AIChE had been pursuing for many years, for example, encouraging changes in educational programs to better prepare chemical engineers for professional work and intensifying AIChE technical activities to meet the evolving needs of the profession. Others called for the continuation of programs begun in the post-1960 era. For example, Objective 1 called for providing technical information to the public and government; Objective 5 called for the enlargement of continuing education programs; and Objective 7 called for promotion of employment practices mutually benefiting employees and employers. A few of the objectives pointed in new directions. For instance, Objective 3 proposed formulation of a position on licensing and certification and the development of programs to meet members needs in these areas, and Objective 9 summoned the AIChE to strive for a balance in the supply and demand of chemical engineers.

In an attempt to insure that the objectives were not quickly forgotten, a subcommittee headed by William Corcoran developed an "Action Plan for Dynamic Objectives in Chemical Engineering," issued by AIChE in August, 1978. This document, along with a matrix chart, indicated the responsibilities of each national committee, local section, and division in implementing the objectives. For several years after this each Vice President took responsibility for checking annual progress in meeting the objectives.[151]

Although a form of long-range planning, the 1977 Dynamic Objectives were primarily geared toward internal Institute problems, some very immediate. They were not specifically geared to the long range. Thus in April, 1980, W. R. Marshall, Jr., former President and, at the time, Treasurer, of the AIChE, concerned that another crisis similar to that of 1970-72 might be approaching, again brought up the need for regular long-range planning. He pointed out that the meetings of both the Executive Committee and Council tended to be taken up entirely with matters of immediate concern, with long-range trends all but completely neglected.[152] As a result of this concern, the Executive Committee held a two-day meeting in July, 1980, devoted entirely to long-range planning, and in November, 1980, Council voted to establish a standing committee devoted specifically to long-range planning.[153]

Reorganization Plans

Council's preoccupation with immediately-pressing Institute business to the neglect of broader considerations had been recognized before Marshall. In 1971, in an effort to give Council members an opportunity to consider informally long-range plans and broad ideas,

Martin, then AIChE President, began holding Friday evening brainstorming sessions preceding regular Saturday Council meetings.

One of the ideas emerging out of the August 27, 1971, brainstorming session was the need to consider restructuring the Institute so that it could react more quickly to problems encountered by the profession such as the employment crisis it was then in the midst of. Council voted at its meeting the following day to appoint an *ad hoc* committee to study possible reorganization of the Institute.[154] In October the Institute's Executive Committee asked R. B. Filbert (Director 1971–73) to chair this committee.[155]

When it began meeting extensively in early 1973, Filbert's committee reviewed a large number of organizational plans and suggestions, among them electing national officers with an electoral college system; regionalizing the Institute; enlarging Council to include divisional, committee, and local section delegates; creating a bicameral governing structure; and developing a dues structure independent of membership grade.[156] Discussion centered most heavily, however, around a proposal to create Council positions restricted to members under the age of 35. This proposal, originating in the Rochester Section, had been discussed by Council in March and November of 1973, before being referred to the Filbert Committee.[157] Supporters of the idea based their arguments on a study of AIChE age distribution compiled by John Friedly and Herman Osmers.[158] This study indicated that most of the AIChE members who dropped membership were young: the median age was about 30 and there was a sharp peak at age 27. Further, it revealed that while the median age of AIChE members was 36 years, that of AIChE Directors was 47, and that members under 35 practically never held national office. Proponents of the young director proposal argued that the AIChE lost a large portion of its younger members because it did not seem responsive to their needs and ambitions, something an expansion of Council to include young directors would correct.

Filbert, in his report to Council in August, 1974, noted that on almost every question of reorganization there was considerable polarization within his committee and that there was no consensus on any major reorganizational plan.[159] Only four suggestions received even a high degree of committee support:

- creating a dues structure independent of grade of membership but dependent on age so that young graduates would have lower dues;
- creating an organization of local section officers with the autonomy of a division and direct access to Council;
- developing procedures for identifying and involving younger members in national AIChE activities; and
- requiring the Nominating Committee to select candidates "to achieve the maximum degree of equitable representation on Council with regard to consideration of geographical, age, and type of employment characteristics of the membership."[160]

These recommendations suffered varying fates. One was completely rejected. Council feared that a "pro-Council" of local section officers would have little continuity from year to year, that many local section officers would have neither the time nor the money to attend the meetings of the proposed body, and that its creation might lead to divided authority within the AIChE. Two of the recommendations were partially adopted. The national AIChE did increase its efforts to bring younger members into national activities, but set up no formal system, like the young directors plan, to do this. Similarly, the Nominating Committee was advised to strive for "equitable representation" for different geographical regions, age groups, and employment groups, but its formal charge was not altered.[161] One of the committee's recommendations was eventually adopted in full. In April, 1976, Council reversed its previous stand and adopted a dues system based on age rather than membership grade.[162]

Several things explain why the internal organization of the Institute was not further altered during the 1970s. First, the lack of consensus within the organization on the direction reorganization should take made it easier to stay within the existing structure. Second, none of the major reorganizational suggestions really offered any radical improvement in performance over the existing organization. Third, as the recession faded, so did the support for major organizational changes. The political apathy of prosperous times replaced the political activism of troubled times before any modifications had been made. Finally, the Institute's attempts to meet the demands of its younger members through expanded employment services, the development of employment guidelines, and economic surveys helped quell the charge that the Institute needed restructuring because it was not meeting the needs of its members.

Modified Membership Requirements

Although the Institute was not massively restructured as a result of the agitation of the early 1970s, the concerns which had led to the Filbert Committee did bring some changes. Among the most important were easier procedures for transferring from the Associate Member to the Member grade. This change was seen as an alternative to the young director proposal as a means of more fully involving younger members in AIChE affairs.

In the early 1970s approximately two-thirds of the Institute's members, including virtually all of its younger members, were Associate Members, a grade restricted by the constitution from holding national office or voting on constitutional amendments.[163] In 1975 an *ad hoc* committee on membership grades, chaired by Timmerhaus, proposed eliminating the grade of Associate Member and making its criteria the criteria for election as a full Member.[164] Council, concerned about the heavy dropout rate among younger AIChE members, and wishing to involve them more fully in Institute affairs, strongly favored the suggestion.[165] But nothing came of it. Informal surveys of full Members, those who would have to approve a constitutional amendment making the appropriate changes, indicated that the 75% favorable vote needed for a constitutional change to lower barriers for full membership could not be attained.[166]

Thus Council, in August, 1976, abandoned the proposal to eliminate the Associate grade and instead proposed modifications to the constitution that would make it easier for Associates to transfer to the Member grade. The AIChE constitution had required, for members with a B. S. degree in chemical engineering, six years experience in chemical engineering practice, including three years in "responsible charge" of chemical engineering work. Council proposed deleting the term "responsible charge" (always hard to interpret) from the requirements and requiring four instead of six years of experience.[167] Constitutional amendments incorporating these changes were approved by an 84% majority in early 1977.[168]

In 1982 Council again seriously considered abolition of Associate Membership, but, again, decided instead to ease the procedures for upgrading membership from the Associate to the Member grade. Under the new system, approved by Council in June, 1982, upgrading would be nearly automatic, dependent on a self-declaration of education and experience in chemical engineering in response to a letter sent by headquarters to members who had been in the Associate grade for the requisite number of years.[169]

Elections

Another area of internal organization influenced by the ferment which emerged from the employment recession of 1970–72 was electioneering. Traditionally, as we have seen, the Institute had encouraged competitive elections for directorships and for vice president, but had discouraged the presentation of specific election platforms. Occasional objections to this practice had emerged prior to 1970, but so long as there was no major divisive issue before the AIChE, such as employer-employee relations, the level of discontent with campaigns based mainly on candidates' professional and service records was not high.

The recession and the rise of discontent among substantial elements of the Institute over Council's moderate approach to employment problems changed this. For example, at the 1970 local section officers' meeting there were complaints about the absence of platforms giving voters an idea about where candidates stood in the biographical sketches published in *CEP* and sent out with election ballots.[170] And in March, 1971, the Central Pennsylvania Section noted with "concern and disapproval" the national organization's policy of prohibiting candidates for national office from presenting a campaign platform.[171]

In May, 1971, responding to this pressure, Council modified its position, permitting the write-up in *CEP* to cover a candidate's areas of interest and concern, problems which the candidate believed the profession faced, and the candidate's views of what the general objectives of the profession and the Institute should be, as well as his professional background. Council, however, maintained its opposition to highly specific political platforms.[172]

The competitive nature of AIChE elections, which had caused occasional problems in the 1960s, continued to cause occasional problems in the 1970s. A special ad hoc committee appointed in January, 1977, and chaired by S. E. Isakoff (Director, 1977–79), revised AIChE's 1969 electioneering guidelines. The new guidelines required that campaigns be financed by the candidates themselves and not by their organizations; prohibited the use of the funds or resources of the Institute, its local sections, or its divisions in elections; and reaffirmed the Institute's prohibition of the mass distribution of printed campaign literature. These guidelines were approved by Council in June, 1977.[173] In February, 1978, they were amended by specifically requiring election campaigns to be "inexpensive" and by defining the term to mean $200 to $300.[174] This specific limit was removed in 1981.[175] After a thorough study by the Constitution and Bylaws Committee, Council in 1982 approved and sent for letter ballot to Fellows and Members changes regarding election of officers. These were approved by a 93% vote and formalized the newly elected Vice President as President elect, combined the corporate Secretary and Executive Director positions, and provided for a contested election for Treasurer for a three year term.

Of course, not all of the changes in AIChE's internal structure between 1970 and the organization's 75th anniversary were the offspring of the 1970–72 employment recession. A number of improvements in the organization of national headquarters and the Institute's local sections and divisions were stimulated by other factors such as the need to improve efficiency in a larger organization.

National Headquarters

One of the major changes in national headquarters between 1970 and 1983 was the reorganization of staff along functional lines. The growth of headquarters staff and staff responsibilities between 1940 and the late 1970s took place without any formal reorganization, which is often the case with a growing organization. As a result, often one person in national headquarters was charged with responsibility for divergent and seemingly unrelated functions, making it difficult for members not intimately familiar with headquarters to know whom to contact about specific problems. One of Forman's first actions after becoming the Institute's Executive Director in mid 1978 was to reorganize headquarters staff into five functional areas—educational services, member activities & services, finance & office services, professional services, and publications and technical services—each headed by a specific Staff Director. This arrangement more clearly identified responsibilities within AIChE and grouped related responsibilities in the same department.[176] In the Diamond Jubilee Year, the senior staff included three staff directors who had served the Institute for many years: Larry Resen, Publications & Technical Services; Hal Abramson, Educational Services; and Harold Hansen, Finance & Office Services & Controller. Following the retirement of Sylvia Fourdrinier and Joel Henry, Gloria Lambson became Staff Director Member Activities & Services and Gerard Chiffriller became Staff Director Meetings & Expositions.

During the late 1970s and early 1980s Forman also started a major upgrade of computer equipment for recordkeeping. By 1983 practically all membership records and files had been converted to a central computer-terminal system with entry from desk top terminals.[177]

The national AIChE's relations with its two primary sub-divisions—local sections and divisions—followed divergent paths in the 1970s. Ties between national and local sections were generally strengthened; ties between national and the divisions, at least for a time, weakened.

Local Sections

Relations with some local sections were somewhat strained by the recession and by differences of opinion on employment guidelines implementation. But despite occasional problems, the ties between the national AIChE and its local sections grew in the 1970s. This was, in part, due to the continuation of the Institute's dues rebate program. But other things contributed. For instance, in May, 1972, Council liberalized the prohibition against non-AIChE members in local sections by permitting members of the Founder Societies to join AIChE local sections without joining AIChE.[178] Even more important was the initiation in 1972 of procedures which permitted dues for local sections to be collected with national dues bills. By 1974, 65 local sections were having their dues collected through national office billing.[179] This practice significantly strengthened a number of local sections. By October, 1977, the gain in local section membership attributable to dues collection by national varied from 10% to 393% and averaged 128% (unweighted). National collection of local section dues had become so advantageous that by that date only seven local sections continued to bill for their own dues.[180] By 1982 all U.S. local sections relied on national for dues collection.

The continuation of the annual local section officers' training and orientation program also contributed to the increased ties between the national AIChE and its local sections. The Institute's contribution to this program was increased when, in 1971, during the employment recession, it offered to underwrite officers' fares to the meeting if employers or sections could not pay them.[181] This policy was continued in subsequent years,[182] although increased costs placed the program under review beginning in 1979.[183] The conferences themselves remained highly popular, especially when expanded to include divisional officers and committee chairmen. Of 56 officers responding to a questionnaire on the 1980 conference, 23 rated it excellent, 33 good, 0 fair or poor, and a number of officers indicated that the meeting was one of the best services offered by the organization.[184] Much of the high popularity of the conferences after 1979 can be attributed to a change in format. One of Forman's innovations was to replace formal staff presentations with workshops involving more participation on the part of attendees.

While the Institute increased its services to local sections, it also imposed increased responsibilities on those sections. In June, 1972, for instance, the Executive Committee advised them that if they did not deliver an annual report on activities to the Institute, rebate checks would be withheld.[185] And in November, 1974, Council accepted a report from an *ad hoc* committee chaired by A. S. West detailing guidelines and goals for local sections and providing the Institute with criteria on which to judge their performance and continued vitality.[186]

Divisions

The Institute's local section structure had largely matured and stabilized by 1970; its divisional structure was still in a state of rapid growth. Between 1970 and 1983 the AIChE expanded from 7 to 11 divisions. A Computing & Systems Technology Division emerged in 1977 out of the old Machine Computation Committee. A Safety & Health Division was formed in 1979, and Marketing and Management divisions were organized in 1980. In addition to assuming programming responsibilities and functioning as autonomous political units in AIChE, most of the divisions provided other services for their membership. Practically all issued directories and newsletters. In the 1970s most created divisional awards to honor contributions to their specialties. And several of the divisions—Environmental, Heat Transfer, Nuclear Engineering, Forest Products, and Food, Pharmaceutical & Bioengineering—published almost an annual volume in the Institute's *Symposium Series*.

In the period of rapid local section growth (1940-60), the Institute had often neglected relations with its local sections and, as a result, had faced the problem, in the early 1960s, of bringing groups which considered themselves almost independent societies back into the AIChE. By the late 1970s there were indications that a similar problem was emerging with divisions. Ties between divisions and the society were often weak, limited to dues collection, divisional officers' meetings held in conjunction with local section officers' meetings, and arrangement of programming space at national meetings. Bylaws requiring divisions to submit an annual report of activities were largely ignored by both national headquarters and divisions.[187] As a result divisions developed largely on their own after establishment. In 1981 the Institute's new Long-Range Planning Committee warned that as divisional activities and autonomy increased many divisions had begun to think of themselves as "technical societies in their own right."[188]

This situation was of growing concern to the Institute's leadership since in some societies divisions had become so strong that virtually all activities emerged from the technical divisions with "little credence" being given to suggestions from national officers.[189] One of the thrusts of Forman's early years as Executive Director was to strengthen ties with the organization's divisions. For instance, in 1980 the Institute began to enforce the bylaws which required submission of an annual activities report by divisions. In 1981, at its summer Detroit meeting, the national AIChE sponsored the first meeting of divisional chairmen to discuss common divisional problems. Finally, and most important, the Institute significantly expanded its publications program (reviewed later in this chapter) to meet the needs of several of its larger divisions.

TECHNICAL PROGRAMS

In the early 1970s, with the exception of the continuing education block of the meetings program, AIChE's programs in the technical area (meetings, publications, research, standards) grew very little. The recession of 1970–72 was primarily responsible for this. A decline in advertising revenue accompanying the recession limited revenue growth and thus program growth generally. Membership, reflecting unemployment and disenchantment with professional societies, stagnated at around the 36,000 mark (for dues paying members) between 1970 and 1975. This stagnation, naturally, did not encourage the enlargement of existing technical programs. Moreover, the unemployment crisis put pressure on the Institute to sacrifice or hold the line on its technical programs in order to expand programs focusing on the economic needs of its membership.

In the late 1970s and early 1980s, as the profession recovered from the 1970–72 recession and as the demands for economic programs declined, the Institute once again began to expand its traditional technical programs in meetings and publications. Moreover, the AIChE also revived, in the mid 1970s, its involvement in research, an activity it had abandoned in the 1960s.

Meetings

The regular meetings program grew slowly in the 1970s and early 1980s. The number of sessions, for instance, increased from an average of around 225 annually to only around 250 to 275 annually. Yearly attendance grew hardly at all. There were substantive reasons for these developments. First, beginning in 1975 the number of annual technical meetings was reduced from four to three, partially because of fears that the Petrochemical and Refining Exposition (held in odd-numbered years) would attract so many members that the technical meeting which followed would be too sparsely attended and partially because of growing requests for AIChE sponsorship of specialized meetings with other technical societies and international organizations. Second, the growth of technical sessions at meetings was increasingly restricted by the limitations of convention facilities. In addition, in 1982 the Institute deliberately imposed limits on the length of meetings and the number of sessions which could be run simultaneously in hopes of improving the overall quality of papers delivered.[190]

The two primary programs held in conjunction with AIChE meetings—industrial expositions and continuing education—enjoyed greater growth in the 1970s than the technical meetings program.

The AIChE Petrochemical and Refining Exposition, initiated in 1961 and held in conjunction with the first technical meeting in odd-numbered years, continued to prosper in the 1970s. Registration at the Petrochemical Exposition hit a temporary plateau at around 15,000 in 1969 and 1971 and then declined sharply in 1973, reflecting the economic depression of the early 1970s and the limitations of the New Orleans facility. From 1975 on, however, all of the Petrochemical Expositions were held in Houston and, due to both the greater popularity of the Houston location and better economic conditions, registration again began to grow, exceeding the 30,000 mark by 1977.

Encouraged by both the popularity of the Petrochemical and Refining Exposition (commonly called Petro Expo) and the revenues it brought to the Institute, the AIChE initiated in the late 1970s an equipment exposition for even-numbered years: the Chemical Plant Equipment Exposition. The first CPE show, held in Philadelphia in June, 1978, attracted over 300 exhibitors and around 9000 registrants.[191] The 1980 CPE Exposition did not attract a larger crowd, prompting the Institute to move it to the West coast for 1982.[192] In 1981, Forman brought management of AIChE's shows in-house.

The growth of the continuing education program paralleled the growth of AIChE-sponsored industrial expositions. Already, by 1970, the Institute's continuing education program was one of the largest among engineering societies, offering that year 48 courses with 1285 in attendance.[193] Under the direction of Hal Abramson, the program continued to grow in the 1970s. The existing package of Today Series courses, Advanced Seminars, Management Seminars, and workshops was expanded to include occasional seminars designed for high level executives in the chemical industry and a modular instruction series designed primarily for self-study. At the same time attempts were made to fill in gaps in the Institute's offerings, to provide a complete curriculum, and to improve quality control. By the late 1970s the program was running so successfully that the Institute's Continuing Education Committee was no longer spending much time soliciting new course offerings, but was, instead, spending much of its time reviewing unsolicited proposals for new course offerings.[194] In 1981 the AIChE ran 256 courses with an attendance of 4072.[195] However, the 1982 recession adversely affected the continuing education program; attendance fell to 2805.

By 1980 the American Institute of Chemical Engineers was an acknowledged leader among major engineering societies in the area of continuing education. The eminence of its position can perhaps best be judged by comparing its program to those of other major professional engineering and scientific societies. In 1976, for instance, the AIChE allotted $8.26 per member of its budget for educational programs, nearly twice the amount allocated by any of the other Founder Societies. The ASCE allocated $4.84, the IEEE $2.22, the AIME $1.78, and the ASME $1.34.[196] Moreover, in 1981 AIChE continuing education courses had 81 registrants per 1000 members, more than five times the proportion of any of the other Founder Societies.[197]

In conjunction with its continuing education program, the AIChE initiated in 1979 a Professional Development Recognition Certificate (PDRC) program. This program was created in response to fears prompted by Iowa's passage in 1978 of the first American legislation requiring licensed engineers to provide proof of continuing professional competence when renewing their licenses.[198] The Professional Development Recognition Certificate was intended to do this. It required documentation of certain levels of activity in three areas: (1) chemical engineering practice, (2) informal educational activities such as self-study and volunteer service to technical and professional societies, and (3) formal educational activities such as attending continuing education courses, attending technical meetings, and presenting papers.[199] At the AIChE's 75th anniversary the future of the certification program was still in doubt, its future in large part dependent on whether other states chose to imitate Iowa's legislation.

The increased number of sessions held at the more popular AIChE meetings and the expansion of the Institute's continuing education and exposition programs began in the 1970s to place some strains on program organization. Responsibility for organizing meetings was divided three ways. Divisions and specialist committees of the national Program Committee had responsibility for much of the programming. They organized sessions in their specialties and secured speakers for these sessions. The local sections at particular meeting locations had responsibility for arranging plant trips, organizing social programs, and handling registration at the meetings. National headquarters working with the Meeting Program Chairman and the Executive Board of the Program Committee had responsibility for determining the location of the meetings, selecting convention facilities, making hotel reservations, organizing the technical program to fit space and time limitations, publishing and distributing the program, and so on. In 1980 this system required modification. National headquarters took over all responsibility for registration since some small local sections did not have sufficient manpower to handle it effectively.[200] Moreover, the Institute improved registration efficiency through the increased use of mini-computers, both in headquarters and at meeting sites.

The Institute also made other adjustments to ease strains within or between the different groups responsible for programming. In 1980, for instance, the national Program Committee was reorganized into two broad programming areas: fundamentals and applications. The programming groups in these areas were arranged in a matrix to indicate opportunities for joint programming between groups in each area.[201]

Divisions and program groups in the national Program Committee complained with increasing frequency in the 1970s about the restrictions placed on the number of sessions they could schedule. In 1980 the Executive Board of the Program Committee suggested meeting these complaints by extending some meetings an extra half day, by attempting to shift some programming to less heavily used meetings (particularly the summer meeting), and by encouraging "block" programming by divisions in conjunction with regular AIChE meetings. AIChE's leadership hoped that the policy of encouraging divisions to concentrate a large portion of the sessions they usually sponsored through the course of a year into a "block" presented at a single meeting would increase meeting attendance, encourage interaction between AIChE members in various specialties, and, thus, help offset the centrifugal tendencies of completely independent specialty meetings.[202]

Publications

AIChE's publications program was closely linked to the continued vitality of AIChE's meetings program since AIChE meetings continued to be the primary source of papers for the society's publications between 1970 and 1983. In the 1970s the Institute maintained the level of publication activity attained in the 1960s in spite of occasional pressure to reduce publications and funnel more funds into non-technical programs. A survey of several groups of AIChE members in the mid 1970s indicated that "a large majority" continued to be satisfied with AIChE's major publications (*CEP, AIChE Journal,* the *Symposium Series,* the *Technical Manuals,* and *International Chemical Engineering*) and would omit "none of them."[203]

In the late 1970s, however, the Institute's publications program faced challenges from several directions. The steadily increasing number of papers being presented at AIChE meetings (around 1200 annually, not counting papers presented at specialized meetings) placed pressure on the system.[204] In addition, a number of other organizations, publications, and groups had begun to expand their publications programs into areas traditionally in the domain of the chemical engineer. Concerned about these developments, Council in August, 1979, at the suggestion of William Corcoran, then serving as Past President, authorized a Publications Task Force of the Publication Committee to review the AIChE program.[205]

The Publications Task Force concluded that current AIChE publications were doing an "adequate job" and that there was "nothing inherently wrong" with AIChE's program. The committee, however, recommended improvement and expansion within the existing structure through serialization of segments of the irregularly-published *Symposium Series* and the *Technical Manuals.*[206] In November, 1980, at the recommendation of the Publications Task Force, Council approved creation of three new quarterly AIChE publications: *Energy Progress, Environmental Progress,* and *Plant/Operations Progress.*[207]

These three new quarterlies were designed to absorb many of the papers that would normally have been published in AIChE's *Symposium Series* and *Technical Manuals,* but with significant improvements. Because *Energy Progress, Environmental Progress,* and *Plant/Operations Progress* were quarterlies, publication would be more rapid and more frequent than with the *Symposium Series* or *Technical Manual* volumes, which often came out only on an annual basis. *Symposium Series* volumes had varied in quality, depending on who served as editor. The new quarterlies were each to have individual editors and much tighter editorial liaison with national headquarters and, hence, higher overall quality. The expanded program also offered a superior outlet for the publication ambitions of several important AIChE divisions.

The Institute hoped that by initiating the new quarterlies on existing foundations the transition would be smoother and the period before operations became self-sustaining would be shorter. These hopes were immeasurably aided when both the Fuels & Petrochemical Divison and the Environmental Division voted to raise their dues beginning in 1982 in order to have their members receive, respectively, *Energy Progress* and *Environmental Progress.*[208]

Research

Traditionally, the AIChE's technical program had focused on meetings and publications. Attempts to add a strong third element to these through initiation of substantial involvement in standards activities or through sponsorship of research had failed in the 1950s and 1960s. The Institute did become involved in setting standards in the nuclear field through the ASA, later ANSI, but standards never did become a major AIChE program. Research was abandoned in the mid 1960s after two only modestly successful endeavors: the bubble tray efficiency project and the machine computation of physical properties project (reviewed in the preceding chapter).

In the early 1970s, however, the AIChE reversed itself and revived its research program. The factors behind this decision are not completely clear. To some extent it may simply have been a matter of different personnel with different interests on Council. But other factors contributed to the decision. For instance, in the early 1960s, when AIChE abandoned its research program, the federal government's involvement in subsidizing scientific and engineering research was reaching its peak. Both universities and private research facilities were having little trouble securing research support, minimizing the need for an AIChE role in the area. By the 1970s, however, the federal government had significantly cut back on its subsidization of research, leaving a vacuum that AIChE could partially fill. The failure of the Industrial Associates Plan may also have provided a stimulus for taking another look at the question of raising funds from industry to support research. One of the factors behind AIChE's decision to abandon fund-raising for research on a case-by-case basis in the early 1960s was the Institute's hope that industry, through the Industrial Associates Plan, would provide substantial support for the general AIChE budget on an annual dues-paying basis.

The Executive Committee first gave serious consideration to reviving AIChE sponsorship of research in early 1968 in considering a proposal from A. E. Dukler and Ovid Baker. They suggested the Institute sponsor a committee which would explore industry's need for specific design data on two-phase flow and give that committee the right to raise funds and convert itself into a research organization if it seemed advisable. In support of this proposal, they pointed out that research on two-phase flow, previously sponsored by the American Gas Association and the American Petroleum Institute, had produced design data for pipelines only. The field as a whole remained in a state of "disarray" with no one trying to relate the general and highly theoretical university research on two-phase flow to the very specific design needs of industry.[209]

Both the two-phase project and the entire question of the Institute's role in raising funds for research were considered at subsequent Executive Committee and Council meetings in 1968 and 1969. In August, 1969, Council reversed the position the Institute had maintained since 1964 by approving a revision of Research Committee policy permitting, with prior Council approval, "the solicitation of funds or sponsors for approved [research] projects."[210]

In November, 1970, Council reviewed a proposal for the creation of a Design Institute for Multiphase Processing (DIMP) presented by Baker and Dukler,[211] and in February, 1971, Council approved "in principle" the concept of creating such an organization, requesting the Research Committee to work out a detailed program for implementation.[212]

The basic idea behind the design institute was similar to the basic idea behind AIChE's earlier research programs: to make use of the Institute's services and prestige to promote cooperative research in critical areas where companies were unwilling or unable to undertake research individually. But there were significant differences between the proposed design institute and the earlier bubble tray efficiency and computer computation projects. In the earlier projects AIChE had played a direct role in soliciting funds and administering research. DIMP was to be a semi-autonomous organization. Through its own administrator/promoter it would take charge of fund raising and through its own committees it would direct, administer, and monitor the agencies contracted to carry out the research. AIChE's direct role would be restricted to providing overall guidance, to lending its sponsorship and prestige to the design institute, to contributing seed money to defray administrative expenses, and to providing promotional impetus during the first years of operation. In addition, after work was completed the Institute would publish, after a period, the resulting design manuals.

In August, 1971, Council formally approved the formation of the Design Institute for Multiphase Processing (DIMP), authorizing $10,000 in seed money to get the project started. The projected budget of the new organization was $100,000 for the first year of operation, $200,000 for the following four years.[213] J. H. Rushton (President, 1957) volunteered to serve as project administrator. Solicitation of funds from industry began shortly afterwards.

DIMP encountered early problems. Due to the 1970-72 recession industrial response was poor. Despite contacting 102 companies, some of them several times, Rushton by February, 1973, had secured only about a dozen industrial sponsors and only $65,000 in annual pledges (plus a $25,000 grant from the Engineering Foundation).[214] Despite this, Council in March, 1973, authorized initiation of the research program. Contracts were let out to four universities for the production of practical design manuals covering specific problems in multiphase flow.[215]

DIMP ultimately attracted 17 industrial sponsors who contributed around $100,000 annually to its budget. Since its projected budget for years two through five had been $200,000, the scope of the project had to be scaled back and only liquid-gas two-phase systems were investigated.[216] Despite this, DIMP was relatively successful. By the end of its fifth year of operation it had repaid the $10,000 seed money contributed by AIChE, and 80% of its industrial sponsors had indicated interest in continuing to support research in the area. Thus Council in June, 1978, approved continuation for three more years and, later, approved an additional two.[217] Although the annual contributions by industry dropped after the initial five-year contracts expired, by 1981 DIMP was again operating with an annual budget of around $100,000.[218]

In 1976, as it became clear that DIMP, despite its reduced budget, was operating well and that the design institute approach to research management was less objectionable than the earlier program, H. S. Kemp (Director, 1981-83), Chairman of DIMP's Executive Committee, and J. H. Rushton, its administrator, proposed that AIChE sponsor another project: a Design Institute for Emergency Relief Systems (DIERS). DIERS was to use funds raised from chemical and petroleum companies and engineering contractors to produce manuals that would enable designers to size the emergency relief systems for hazardous fluids. In August, 1976, Council endorsed the idea, authorizing an *ad hoc* committee to prepare detailed proposals.[219]

The final proposal was submitted to Council in March, 1977. DIERS, like DIMP, was to be able to draw on the AIChE for up to $10,000 annually for start-up and for the travel and other expenses of its volunteer business manager/promoter. It was to have a four-year life, with a projected annual budget of $100,000 for the first year and $200,000 annually thereafter. It differed from DIMP primarily in proposing to use commercial industrial laboratories rather than universities to carry out its research work.[220] Financially, DIERS proved more attractive than DIMP. It was able to initiate research by the fall of 1978. By November of 1979, after approximately one year of operation, it had an annual budget of $327,000 contributed by 26 companies.[221]

A third AIChE-sponsored design institute—the Design Institute for Physical Property Data (DIPPR) was proposed in 1977, was authorized in 1978, and initiated work in 1980.[222] DIPPR was formed to provide industry with more accurate and complete physical property data so that chemical processing equipment could be designed to meet the more stringent environmental and safety regulations and the higher energy costs of the late twentieth century.

By the early 1980s three AIChE-sponsored design institutes were in operation and several more were under consideration. AIChE's research program, seemingly dead and buried in the late 1960s, had been born again in the 1970s and by the 1980s was in the process of becoming a strong third element, with meetings and publications, of AIChE's technical program.

Summary

Although the most striking feature of AIChE activities in the post-1970 era was the emergence of significant non-technical programs, the AIChE clearly did not neglect its traditional commitment to a strong technical program. The improvements made in the organization of the headquarters staff and meetings, the establishment of the Chemical Plant Equipment Exposition, in-house show management, the initiation of three new publications, the revival of AIChE involvement in research, and the revitalization of the society's equipment testing procedures program (mentioned in the preceding chapter) provide evidence that the organization continued to make a strong technical base its first priority, even while undertaking significantly increased activities in the non-technical realm.

The efforts made by AIChE's leadership to maintain growth and vitality in the traditional technical area, despite increased activities in non-traditional fields, was as critical in maintaining balance between the organization's various interest groups as its moderate policies in the areas of guideline implementation and governmental interaction. The AIChE, like other technical societies, was under pressure in the 1970s to reduce activities in traditional areas like publications in favor of increased political and economic activism. Had the Institute shifted its activities too strongly to meet these pressures, it might have gained temporary advantages at the cost of its long term health. For instance, a serious decline in the organization's publication activities could have alienated significant segments of major interest groups in the organization. Furthermore, the concern of the typical chemical engineer with economic and social programs is often transient. As the economy improves or governmental policies change, the demand for professional society involvement in economic and political matters declines and interest shifts back to the traditional core activities: publications and meetings. By remaining vital and active in these core areas, despite pressure to sacrifice technical programs and shift resources to non-technical projects, the AIChE was in a good position to maintain, capture, or recapture membership when interest in political and economic activism receded. Thus, the way AIChE's leadership balanced the needs of its traditional technical programs with the demands for non-technical programs in the early and mid 1970s was at least in part responsible for the rapid AIChE growth rate between 1977 and 1983, when membership increased from around 40,000 to around 55,000.

Chapter 4: Notes

1. "Manpower Demand Hits 10-Year Low in 1970," *CEP*, v. 67 (May, 1971) pp. 38–39; also *CEP*, v. 66 (December, 1970) p. 100.
2. *CEP*, v. 65 (June, 1969) p. 14.

3. Personal correspondence, F. J. Van Antwerpen to author, March 22, 1982.
4. The unpublished reports of the "AIChE Annual Economic Surveys" for 1974–1981 indicate that less than 1% of all AIChE members were unemployed in this period.
5. "Report on Annual AIChE Economic Survey," March, 1971, unpublished; Council, *Minutes*, February 27, 1971, p. 21; *CEP*, v. 67 (March, 1971) p. 1.
6. Council, *Minutes*, May 15, 1971, p. 187.
7. *CEP*, v. 67 (October, 1971) pp. 36–37.
8. "Societies Still Failing," *Industrial Research*, v. 16 (January, 1974) p. 85.
9. K. M. Reese, "Professionalism and Technical Societies," *Chemical and Engineering News*, v. 49 (December 13, 1971) pp. 25–26, 28.
10. *Ibid.*, p. 26; Ellis Rubinstein, 'IEEE and the 'Founders'—II," *IEEE Spectrum*, v. 13 (June, 1976) p. 71.
11. Reese, "Professionalism and Technical Societies," p. 26; Sinclair, *ASME*, pp. 209–13.
12. Executive Committee, *Minutes*, July 26, 1966, p. 226, and September 15, 1967, p. 521; *CEP*, v. 63 (October, 1967) p. 30. In 1978 high-limit accidental death and dismemberment coverage was also offered (Executive Committee, *Minutes*, April 30, 1978, p. 102).
13. Roland A. Labine, "Engineers Ask for Action!," *Chemical Engineering*, v. 72 (February 15, 1965) p. 190.
14. For examples of the more aggressive actions desired by some members see *CEP*, v. 67 (August, 1971) pp. 32–33; v. 68 (June, 1972) p. 26; (October, 1972) pp. 6, 12; (November, 1972) p. 43; v. 69 (February, 1973) p. 39; v. 70 (February, 1974) pp. 12, 14; (April, 1974) pp. 8, 12, 37–38; (May, 1974) pp. 38–39, 40–41; v. 72 (August, 1976) p. 7.
15. Executive Committee, *Minutes*, April 12, 1970, p. 116.
16. *CEP*, v. 66 (May, 1970) pp. 14, 16, 17; v. 67 (August, 1971) p. 33; v. 68 (January, 1972) p. 14; (March, 1972) p. 10; (October, 1972) p. 6.
17. *CEP*, v. 68 (March, 1972) p. 10; (May, 1972) pp. 15–16; (July, 1972) p. 38.
18. See note 14 above.
19. *CEP*, v. 68 (March, 1972) pp. 12, 16.
20. Irv Leibson, E. M Jones, and John J. McKetta, "What's the AIChE doing for you?," *CEP*, v. 55 (July, 1959) p. 74.
21. "Attitudes Towards Unions," *CEP*, v. 70 (February, 1974) p. 42.
22. Council, *Minutes*, February 27, 1971, pp. 20–21.
23. James Buchanan, "AIChE Employment Aids Chronology," *CEP*, v. 68 (November, 1972) pp. 45–47, provides a good account of the growth of AIChE's employment service. See also F. J. Van Antwerpen, "A Society's Response to Unemployment," *CEP*, v. 68 (July, 1972) p. 24.
24. *CEP*, v. 68 (March, 1972) pp. 1–2.
25. Executive Committee, *Minutes*, June 24, 1972, p. 270.
26. J. A. Polack, G. E. Montes, and L. B. Smith, "The Professional Earnings and Career Satisfaction of AIChE Members," *CEP*, v. 49 (July, 1953), pp. 17–18, 20, 22.
27. Council, *Minutes*, May 16, 1959, p. 274; Executive Committee, *Minutes*, September 12, 1959, p. 342.
28. Executive Committee, *Minutes*, June 27, 1970, p. 181.
29. Executive Committee, *Minutes*, October 31, 1970, p. 368.
30. Council, *Minutes*, November 28, 1970, p. 411. See also Executive Committee, *Minutes*, October 31, 1970, p. 368.
31. "Report on Annual AIChE Economic Survey," March, 1971, unpublished.
32. *CEP*, v. 67 (October, 1971) pp. 36–37.
33. F. J. Van Antwerpen, "A Society's Response to Unemployment," *CEP*, v. 68 (July, 1972) pp. 23–25; "Report on Annual AIChE Economic Survey," April, 1972, unpublished.
34. Executive Committee, *Minutes*, January 23, 1980, p. 24.
35. *CEP*, v. 68 (April, 1972) p. 2. For more details on the plan see Arthur H. Hale, "The Professional and the Pension Problem," *CEP*, v. 69 (July, 1973) pp. 36–38.

36. Council, *Minutes*, February 27, 1971, p. 17.
37. Council, *Minutes*, February 19, 1972, p. 27.
38. "Happier Retirements Ahead!," *Chemical Engineering*, v. 81 (March 18, 1974) pp. 50–52. For an example of AIChE input into pension discussions in Congress see "Engineers, Pensions, and Congress," *CEP*, v. 69 (May, 1973) pp. 54–56.
39. Council (*Minutes*, June 2, 1973, p. 221) voted to drop out of the PFP plan, but then rescinded the action (*Minutes*, September 8, 1973, p. 379).
40. *CEP*, v. 73 (September, 1977) p. 37.
41. Council, *Minutes*, May 15, 1971, p. 182.
42. Executive Committee, *Minutes*, October 29, 1971, pp. 505, 551–55.
43. Executive Committee, *Minutes*, October 2, 1970, p. 366; October 31, 1970, p. 369.
44. Council, *Minutes*, November 28, 1970, p. 409; Executive Committee, *Minutes*, October 31, 1970, p. 369.
45. Executive Committee, *Minutes*, January 23, 1971, pp. 2, 11–12.
46. Council, *Minutes*, May 15, 1971, pp. 224–25.
47. Council, *Minutes*, November 27, 1971, p. 559.
48. Council, *Minutes*, February 19, 1972, p. 17.
49. "Guidelines to Professional Employment," *CEP*, v. 68 (April, 1972) pp. 23–25.
50. Council, *Minutes*, August 26, 1972, p. 395.
51. *CEP*, v. 68 (November, 1972) p. 43.
52. Don G. Schroeter, "Guidelines Survey Outcome," *CEP*, v. 72 (March, 1976) pp. 25–26.
53. T. W. Tomkowit, "Some Early Feedback on the Guidelines," *CEP*, v. 68 (September, 1972) pp. 50–51; Council, *Minutes*, August 26, 1972, p. 395.
54. "Employer-Employee Guidelines Study," *CEP*, v. 68 (July, 1972) pp. 33–45, for the discussion. See also Council, *Minutes*, February 19, 1972, p. 17.
55. Council, *Minutes*, May 15, 1971, p. 225, and November 27, 1971, p. 559.
56. "15 Societies Endorse Unified Guidelines," *CEP*, v. 69 (June, 1973) p. 124; Irv Leibson, "Why the Unity Document?," *CEP*, v. 70 (April, 1974) pp. 22–23.
57. Executive Committee, *Minutes*, February 10, 1973, p. 6; "Unified Guidelines Endorsed by AIChE," *CEP*, v. 69 (March, 1973) pp. 15–19.
58. "Guidelines to Professional Employment," *CEP*, v. 68 (April, 1972) p. 25; "Unified Guidelines Endorsed by AIChE," *CEP*, v. 69 (March, 1973) pp. 18–19. The AIChE guidelines recommended termination notice of one month plus severance pay of a half month's salary plus two weeks of salary per year of service. The "Joint Guidelines" called for employers to provide notice *or* compensation equal to one month plus one week per year of service.
59. Council, *Minutes*, August 26, 1972, pp. 427–28. For examples of members favoring a more vigorous implementation policy see note 14 above.
60. Eugene Barr, "How Does Industry View the Guidelines," *CEP*, v. 70 (April, 1974) pp. 24–26. See also C. E. Johnson, "Implementing the Guidelines: A Company's Viewpoint," *CEP*, v. 68 (November, 1972) pp. 26–28, 35.
61. Council, *Minutes*, August 26, 1972, pp. 429–32; "Implementing the Guidelines," *CEP*, v. 68 (October, 1972) pp. 12–13; Ted Weaver, "Open Advocate or Subtle Adversary?," *CEP*, v. 69 (December, 1973) p. 26.
62. T. A. Tomkowit, "Some Early Feedback on the Guidelines," *CEP*, v. 68 (September, 1972) p. 50 (290 major companies); Council, *Minutes*, November 10, 1973, p. 513 (10,000 copies).
63. For example: "Implementing the Guidelines," *CEP*, v. 68 (November, 1972) pp. 21–44; T. W. Tomkowit, "Employer-Employee Guidelines Study," *CEP*, v. 68 (July, 1972) pp. 33–45 (with discussion); "Implementation of the Employment Guidelines—Where Do We Go from Here?," *CEP*, v. 70 (April, 1974) pp. 21–38; J. R. Catenacci and Patrick H. McNamara, "Grass-Roots Guidelines Activity," *CEP*, v. 70 (February, 1974) pp. 28–35; J. G. Temple, Jr., "Dow Chemical and the Employment Guidelines," *CEP*, v. 73 (April, 1977) pp. 47–51.
64. Executive Committee, *Minutes*, March 5, 1976, p. 159.

65. Council, *Minutes,* November 10, 1973, p. 513.
66. Executive Committee, *Minutes,* October 26, 1974, p. 480; Council, *Minutes,* March 15, 1975, p. 63; June 8, 1975, p. 177; September 7, 1975, p. 263.
67. For example, *CEP,* v. 70 (February, 1974) p. 14. For the direct opposite point of view (i.e., a resignation of a member because AIChE had become too involved in employer-employee relations) see *CEP,* v. 69 (April, 1973) pp. 10, 12.
68. Dimitrios Tassios, "Should/Can AIChE Do More For Your Economic Welfare," *Chemical Engineering,* v. 78 (April 5, 1971) pp. 106–10; for his comments at the Philadelphia meeting see *CEP,* v. 70 (April, 1974) pp. 37–38.
69. "An Alternative Point of View to 'Open Advocate or Subtle Adversary'," *CEP,* v. 70 (May, 1974) pp. 38–39, and "A Reply to Ted Weaver's Rebuttal," *ibid.,* pp. 40–41. See also G. H. Marcus, W. H. McCarty, T. B. Richey, and H. S. Rossman, "Some Opinions on Employer-Employee Relations," *CEP,* v. 69 (February, 1973) pp. 37–39.
70. Executive Committee, *Minutes,* August 18, 1973, pp. 333–34, 347–50.
71. "Evaluation of AIChE," *CEP,* v. 69 (October, 1973) p. 47.
72. *CEP,* v. 69 (August, 1973) p. 68; (December, 1973) pp. 74–75.
73. *CEP,* v. 70 (August, 1974) p. 27.
74. Council, *Minutes,* November 30, 1974, p. 555; *CEP,* v. 70 (December, 1974) p. 75.
75. Council, *Minutes,* June 1, 1974, pp. 240–41; August 17, 1974, p. 384. See Council, *Minutes,* March 15, 1975, p. 101, for the results of the vote.
76. Council, *Minutes,* November 24, 1979, p. 224.
77. *CEP,* v. 69 (October, 1973) p. 46.
78. Don G. Schroeter, "Guidelines Survey Outcome," *CEP,* v. 72 (March, 1976) pp. 21–33, especially pp. 25–29.
79. Council, *Minutes,* January 24, 1976, pp. 6–8.
80. Council, *Minutes,* November 11, 1978, pp. 264–65.
81. Data from annual "Report on Annual AIChE Economic Survey," 1971–81, unpublished.
82. See Council, *Minutes,* March 31, 1979, p. 49, and August 18, 1979, pp. 157–59, for the initiation of "sunset" legislation. See Executive Committee, *Minutes,* January 16, 1982, pp. 12–13, and Council, *Minutes,* February 27, 1982, p. 45, for the disbanding of the Professional Employment Committee.
83. Herbert Popper, "Questions, Answers, and Conclusions on the Ch. E. and Society," *Chemical Engineering,* v. 79 (October 30, 1972) pp. 111–12.
84. There is a brief review of the growing anti-technology movement of the 1960s and 1970s in Samuel C. Florman, *The Existential Pleasures of Engineering,* New York, 1976, pp. 11–17.
85. Van Antwerpen and Fourdrinier, *High Lights,* pp. 141–44.
86. Council, *Minutes,* September 28, 1963, pp. 340, 435–39; Executive Committee, *Minutes,* June 21, 1969, p. 229; Council, *Minutes,* August 23, 1969, pp. 278–79.
87. *CEP,* v. 67 (September, 1971) p. 43.
88. Council, *Minutes,* May 16, 1970, p. 140; "Environmental Division Set," *CEP,* v. 66 (May, 1970) p. 39.
89. "A Short History of the AIChE Environmental Division," *AIChE Environmental Division Newsletter,* v. 11, no. 3 (September, 1981) p. 4.
90. *Ibid.,* pp. 3, 7.
91. *Ibid.,* pp. 3–4.
92. Council, *Minutes,* August 23, 1969, p. 281.
93. Council, *Minutes,* August 28, 1971, p. 419.
94. Council, *Minutes,* November 27, 1971, pp. 570, 634–38.
95. Council, *Minutes,* February 19, 1972, pp. 26, 70.
96. Council, *Minutes,* November 30, 1974, p. 512.
97. Council, *Minutes.* August 16, 1980, pp. 182–84.
98. *CEP,* v. 67 (February, 1971) p. 12; (May, 1971) pp. 8, 10, 12, 14.

99. John J. McKetta, "Your Greatest Failure," *CEP*, v. 63 (December, 1967) pp. 24–27.
100. *CEP*, v. 64 (May, 1968) pp. 103–04; Council, *Minutes*, February 27, 1971, pp. 77–83; "AIChE Sections Organize to Tackle Pollution Problems," *Chemical Engineering*, v. 78 (April 5, 1971) pp. 68, 70.
101. Council, *Minutes*, February 27, 1971, pp. 65–66.
102. "Pollution Solution via the Grass Roots," *CEP*, v. 45 (April, 1969) p. 27; "Pollution Control," *CEP*, v. 64 (May, 1968) pp. 103–04; "Pollution Study Group," *CEP*, v. 67 (January, 1971) p. 104; "Pollution Solution Reports," *CEP*, v. 68 (January, 1972) pp. 111–13.
103. For example, "Pollution Solution Reports," *CEP*, v. 68 (January, 1972) pp. 111–13.
104. "Short History of Environmental Division," p. 3.
105. Council, *Minutes*, February 27, 1971, pp. 19–20, 64–83.
106. "Short History of Environmental Division," p. 4.
107. *Ibid.*, p. 5; also *CEP*, v. 69 (September, 1973) p. 12.
108. "Short History of Environmental Division," p. 7.
109. Morton Corn, "The Impact of Federal Regulations on Engineers," *CEP*, v. 74 (July, 1978) pp. 24–25; William A. Groening, Jr., "The CPI and the Law: 40 Years Back, 40 Years Ahead," *ibid.*, pp. 17–20.
110. *CEP*, v. 69 (August, 1973), p. 44.
111. Council, *Minutes*, March 10, 1973, p. 170.
112. Council, *Minutes*, November 30, 1974, p. 530.
113. Council, *Minutes*, August 29, 1970, pp. 316–17; March 15, 1975, p. 66.
114. Council, *Minutes*, March 15, 1975, p. 66; November 15, 1975, pp. 391–92, 412–25; January 24, 1976, p. 6.
115. Council, *Minutes*, March 10, 1973, pp. 169–70.
116. Council, *Minutes*, November 30, 1974, pp. 508–09, 528–39; also "AIChE Interaction With the Federal Government," *CEP*, v. 71 (January, 1975) pp. 17–27.
117. Council, *Minutes*, November 30, 1974, pp. 508–09.
118. Executive Committee, *Minutes*, May 9, 1975, p. 167.
119. *CEP*, v. 71 (July, 1975) p. 10.
120. Executive Committee, *Minutes*, May 9, 1975, p. 149.
121. Council, *Minutes*, September 7, 1975, pp. 259–60, 292–99.
122. Council, *Minutes*, January 24, 1976, pp. 3–5, 16–29.
123. "AIChE Interaction With the Federal Government," *CEP*, v. 71 (January, 1975) p. 20.
124. A. S. West, "Engineers in Washington—Good, Bad, or Futile?," *CEP*, v. 74 (January, 1978) pp. 24–25; see also F. J. Van Antwerpen, "AIChE's Washington Involvement," *CEP*, v. 76 (July, 1980) p. 22.
125. Council, *Minutes*, November 12, 1977, pp. 569–70; February 25, 1978, pp. 72–73.
126. Council, *Minutes*, February 25, 1978, pp. 69–76, for the report; quote from p. 73.
127. Council, *Minutes*, February 25, 1978, pp. 20–21; Executive Committee, *Minutes*, April 30, 1978, p. 107.
128. *CEP*, v. 74 (June, 1978) p. 91.
129. Executive Committee, *Minutes*, July 15, 1979, p. 141.
130. Executive Committee, *Minutes*, January 23, 1980, p. 7-8; Council, *Minutes*, March 15, 1980, p. 42.
131. Council, *Minutes*, April 4, 1981, p. 25.
132. Council, *Minutes*, June 7, 1981, p. 126.
133. Council, *Minutes*, June 8, 1975, pp. 177–78.
134. Council, *Minutes*, September 7, 1975, p. 261.
135. Council, *Minutes*, November 15, 1975, p. 396. See also "Northern California Section Opposes Nuclear Initiative," *CEP*, v. 72 (March, 1976) p. 86; "Southern California Section Opposes Nuclear Initiative," *CEP*, v. 72 (April, 1976) p. 95.
136. Council, *Minutes*, January 24, 1976, p. 12 and Appendix I.

137. "AIChE Council Opposes California Initiative," *CEP*, v. 72 (January, 1976) pp. 21–23.
138. "Nuclear-Power Issue Tops Los Angeles AIChE Meeting," *Chemical Engineering*, v. 82 (December 22, 1975) pp. 32–33.
139. "Members Urged to Oppose Anti-Nuclear Initiatives," *CEP*, v. 72 (October, 1976) p. 98.
140. *CEP*, v. 72 (March, 1976) pp. 8, 14, 18; (July, 1976) p. 12.
141. Executive Committee, *Minutes*, April 29, 1977, p. 193; Council, *Minutes*, August 27, 1977, p. 308.
142. Executive Committee, *Minutes*, April 30, 1978, p. 103.
143. Executive Committee, *Minutes*, February 3, 1979, p. 11; July 15, 1979, p. 135.
144. Council, *Minutes*, August 17, 1974, p. 408.
145. The best review of the history of AIChE's long-range planning activities is contained in Executive Committee, *Minutes*, July 9–10, 1980, pp. 153–56.
146. Council, *Minutes*, May 20, 1972, pp. 191–92, 244f.
147. Ted Weaver, "AIChE—Forward Planning for the Years Ahead," *CEP*, v. 69 (January, 1973) pp. 21–26.
148. Irv Leibson, "Delphi Analysis for the Chemical Engineering Profession," *CEP*, v. 70 (March, 1974) pp. 47–54.
149. Executive Committee, *Minutes*, October 26, 1974, p. 484.
150. "Dynamic Objectives for Chemical Engineering, 1977," *CEP*, v. 73 (April, 1977) pp. 41–46.
151. "Action Plan for Dynamic Objectives for Chemical Engineering," American Institute of Chemical Engineers, August, 1978, 16 pp.; "Progress in Implementation of Dynamic Objectives," *CEP*, v. 75 (December, 1979) pp. 73–74; also *CEP*, v. 74 (February, 1978) pp. 82, 84, 88, 96.
152. Executive Committee, *Minutes*, April 26, 1980, p. 87.
153. Council, *Minutes*, November 15, 1980, p. 250.
154. Council, *Minutes*, August 28, 1971, pp. 408–409.
155. Executive Committee, *Minutes*, October 29, 1971, p. 498.
156. Council, *Minutes*, August 17, 1974, pp. 380, 406–19.
157. Council, *Minutes*, March 10, 1973, pp. 167–68, 176–78; November 10, 1973, pp. 510–11.
158. John C. Friedly and Herman R. Osmers, "AIChE Age Profile," *CEP*, v. 70 (December, 1974) pp. 7–14.
159. Council, *Minutes*, August 17, 1974, p. 410.
160. Council, *Minutes*, August 17, 1974, pp. 380, 410–13.
161. Council, *Minutes*, March 15, 1975, pp. 59, 75–79.
162. Council, *Minutes*, April 10, 1976, p. 245.
163. Council, *Minutes*, November 10, 1973, p. 510.
164. Council, *Minutes*, November 15, 1975, pp. 426–41.
165. Council, *Minutes*, November 15, 1975, pp. 393–94.
166. Council, *Minutes*, January 24, 1976, p. 9; Executive Committee, *Minutes*, June 6, 1976, p. 266; Council, *Minutes*, August 28, 1976, pp. 346–47.
167. Council, *Minutes*, August 28, 1976, p. 347; November 27, 1976, pp. 471–72.
168. Executive Committee, *Minutes*, March 24, 1977, p. 144.
169. Council, *Minutes*, February 27, 1982, pp. 42–44; Executive Committee, *Minutes*, April 17, 1982, pp. 70–71; Council, *Minutes*, June 5, 1982, p. 109.
170. Executive Committee, *Minutes*, April 3, 1971, p. 172.
171. Executive Committee, *Minutes*, April 3, 1972, p. 171.
172. Council, *Minutes*, May 15, 1971, pp. 178–79, 226–28.
173. Council, *Minutes*, June 5, 1977, p. 227–29.
174. Council, *Minutes*, February 25, 1978, p. 23.
175. Council, *Minutes*, April 4, 1981.
176. *CEP*, v. 75 (January, 1979) pp. 97–98.

177. *CEP*, v. 78 (October, 1982) p. 138.
178. *CEP*, v. 68 (July, 1972) p. 1; also Council, *Minutes*, May 20, 1972, pp. 195-96.
179. Council, *Minutes*, November 30, 1974, p. 525.
180. Executive Committee, *Minutes*, October 1, 1977, p. 388.
181. Executive Committee, *Minutes*, January 23, 1971, p. 5.
182. Executive Committee, *Minutes*, June 24, 1972, p. 271.
183. Executive Committee, *Minutes*, May 4-5, 1979, p. 75.
184. Council, *Minutes*, June 7, 1980, p. 139.
185. Executive Committee, *Minutes*, June 24, 1972, p. 272.
186. Council, *Minutes*, November 30, 1974, pp. 506, 524-27.
187. Council, *Minutes*, August 16, 1980, p. 180.
188. Council, *Minutes*, August 15, 1981, p. 201; Executive Committee, *Minutes*, July 11-12, 1981, p. 149.
189. Executive Committee, *Minutes*, July 11-12, 1981, p. 149.
190. Council, *Minutes*, February 27, 1982, pp. 40-42.
191. *CEP*, v. 74 (August, 1978) p. 35.
192. Council, *Minutes*, June 7, 1980, p. 135; *CEP*, v. 77 (October, 1981) p. 14.
193. Marshall, "Continuing Education in Chemical Engineering," p. 450.
194. Interview with Harold Abramson, March 25, 1982.
195. *CEP*, v. 78 (October, 1982) p. 21.
196. Ellis Rubinstein, 'IEEE and the Founder Societies," *IEEE Spectrum* v. 13 (May, 1976) p. 78.
197. Harold Abramson, "AIChE Educational Services Department, Report to Council," February 27, 1982 (AIChE records).
198. Council, *Minutes*, June 3, 1978, pp. 144-47.
199. Council, *Minutes*, March 31, 1979, pp. 54-55, plus Appendix A.
200. Executive Committee, *Minutes*, January 23, 1980, p. 20-21; Council, *Minutes*, March 15, 1980, p. 50.
201. *EBPC News* (Newsletter of the Executive Board of the Program Committee), n.d., [late 1980].
202. Executive Committee, *Minutes*, April 26, 1980, pp. 87-88.
203. Council, *Minutes*, September 7, 1975, p. 257.
204. Council, *Minutes*, November 15, 1980, p. 264.
205. Council, *Minutes*, August 18, 1979, p. 163.
206. Council, *Minutes*, March 15, 1980, p. 48; November 15, 1980, pp. 290-309, for the report.
207. Council, *Minutes*, November 15, 1980, pp. 264-65.
208. *CEP*, v. 77 (November, 1981) p. 70; Executive Committee, *Minutes*, October 3, 1981, p. 211.
209. Executive Committee, *Minutes*, January 20, 1968, pp. 6-7; April 27, 1968, pp. 104, 118-21.
210. Council, *Minutes*, August 23, 1969, pp. 277-78, 315-16.
211. Council, *Minutes*, November 28, 1970, pp. 396, 417f.
212. Council, *Minutes*, February 27, 1971, pp. 17-18.
213. Council, *Minutes*, August 28, 1971, pp. 410-11, 425-36.
214. Executive Committee, *Minutes*, February 10, 1973, pp. 1-2.
215. Council, *Minutes*, March 10, 1973, p. 167.
216. Council, *Minutes*, November 15, 1975, pp. 395-96.
217. Council, *Minutes*, June 3, 1978, pp. 132-33.
218. Council, *Minutes*, November 7, 1981, p. 250.
219. Council, *Minutes*, August 28, 1976, pp. 351-52. DIERS was formally approved by Council in March, 1977 (Council, *Minutes*, March 19, 1977, pp. 83-84).
220. Executive Committee, *Minutes*, March 24, 1977, pp. 133-42; Council, *Minutes*, March 19, 1977, p. 83.
221. Council, *Minutes*, November 24, 1979, p. 243.
222. Council, *Minutes*, June 3, 1978, pp. 134-35; November 24, 1979, p. 243; November 15, 1980, pp. 268-69; November 7, 1981, p. 252.

POSTSCRIPT

The American Institute of Chemical Engineers is one of the youngest of the major American engineering societies. When it was founded in 1908 the American Society of Civil Engineers was over half a century old. The American Institute of Mining, Metallurgical and Petroleum Engineers had been in existence for almost four decades and the American Society of Mechanical Engineers and the American Institute of Electrical Engineers for almost three decades. Moreover, as we have seen, the AIChE emerged under far more difficult circumstances than any of these societies. In 1908 the very existence of chemical engineering as a distinct discipline was in question, and the American Chemical Society had already staked out claims to the domain of the infant Institute.

Despite this late and difficult birth, the American Institute of Chemical Engineers has played a role far out of proportion to its age and size. It pioneered in the accreditation of engineering colleges. It was the first major engineering society to use computers for data processing, eventually supplying that service to a number of other societies. In more recent years, AIChE pioneered in the development of mechanical information retrieval for engineering publications, led resistance to the implementation of the ASEE "Goals" report, led the effort to develop a set of joint employment guidelines for engineers, and developed a continuing education program second to none among major engineering societies. Not only is AIChE a member of the prestigious Founder Societies, and hence part owner of the United Engineering Center, but it is both the oldest and the largest chemical engineering society in the world.

The AIChE has also been successful in representing its discipline. By most observers' calculations the AIChE has a greater percentage of the chemical engineers than any other engineering society has of its possible members. Although solid statistical data on this subject are not available, AIChE officials estimate that between 50% and 75% of all practicing chemical engineers are members of AIChE.[1] Comparable figures for other major engineering societies are much lower, around 20% to 40%.[2]

Why has the youngest and smallest of the Founder Societies played such an important role in the engineering profession as a whole and succeeded in attracting and holding such a large portion of its discipline?

In part AIChE's major early contribution to the engineering profession in America (accreditation) was the result of a successful response to a particular set of circumstances not faced by the other major engineering societies. No other major American engineering group was faced with the necessity of justifying the right of its discipline to exist to the same extent as the American Institute of Chemical Engineers.

Another factor contributing to AIChE's success has been the ideology of chemical engineers. Chemical engineers firmly believe that their branch of engineering is more broadly based (because of the inclusion of chemistry, in addition to mathematics and physics) than any other branch of engineering. The pride which this belief has generated is one factor which has caused chemical engineers to rally strongly around the organization claiming to be the spokesman for their profession and to seek leadership positions in the engineering profession as a whole.

While the particular challenges faced by the early AIChE and pride in the uniqueness of chemical engineering have contributed to AIChE's accomplishments, much of its success, in more recent years, has rested on the nature of its governing structure. For example, the AIChE's tradition of open, contested elections and the ease with which candidates can be added to the ballot by petition have strongly contributed to the society's success in retaining membership and operating effectively. These policies have kept interest in AIChE elections at a relatively high level. Candidates winning contested elections have taken their responsibilities more seriously than they might have had elections been uncontested. And the democratic nature of Institute elections has permitted those discontent with prevailing policies an option—appeal to the ballot box—to dropping membership, becoming apathetic, or forming splinter societies.

The nature of the organization's governing body—Council—has also contributed to the growth and health of the Institute. AIChE's Council, while it has varied somewhat in size through the years (currently there are 17 members), has always remained a manageable size. It is large enough to permit the infusion of ideas, but small enough to debate and act on matters effectively. Moreover, unlike similar bodies in other professional societies, AIChE's Council has continued to be elected on an at-large basis. Thus individual Council members are less tied to particular constituencies with particular pet interests than the directors of many other societies. At-large election has tended to produce directors more concerned with the overall health of the organization than with the concerns of particular divisions, geographical regions, or age groups within the profession. This has helped the Institute to maintain balance between its interest groups.

Another important factor behind the vitality of the Institute, particularly in the last quarter century, has been the emergence of the Institute's Executive Secretary (now Executive Director) as a strong executive officer. In the 1950s, AIChE's Executive Secretary took on a dual role. He became both general manager for the society (with vastly increased authority and responsibility) and Secretary (with vote) for Council. The uniqueness and importance of this arrangement was recognized in the 1967 "Future of the Institute" report, which commented: "Other societies do not assign their secretaries this dual role and therefore limit their effectiveness."[3] It has worked well.

Postscript Notes

1. Conversations with J. Charles Forman, Executive Director, AIChE, and Larry Resen, Director of Publications, AIChE. Gloria M. Lambson, "Forecasting Employment Demand for Chemical Engineers," M.B.A. thesis, New York University, 1976, p. 19, estimated that 69.1% of all chemical engineers belonged to the AIChE in 1970.
2. If one compares the number of civil and mechanical engineers reported in census returns with ASCE and ASME membership, these organizations seem to contain around 30–35% of their potential membership. The proportion of registered professional Engineers belonging to the National Society of Professional Engineers is much lower. The same applies to the proportion of manufacturing engineers belonging to the Society of Manufacturing Engineers (see *CEP*, v. 70 [March, 1974] p. 40).
3. "Report of the Ad Hoc Committee on the Future of the Institute," February, 1967, in Executive Committee, *Minutes,* April 14, 1967, p. 155.

CHAPTER 5

Anecdotes

During the planning for the Diamond Jubilee history, the suggestion was made that it include a section containing anecdotes from the 24 then living Presidents who have served since the Golden Jubilee in 1958. The suggestion was well made, for in the following pages we have from them comments as to how they viewed their tenure on the Executive Committee, where they served in the capacity of Vice President, President, and Past President.

The contributions have a wide variety of content: serious, sensitive, perceptive, humorous. There is a degree of repetition in them, and this is left in purposely, for the pages are meant to show what stood out most strongly in their individual recollections.

The section on John Healy, 1961 President, and the only non-living one when this section was organized, was compiled based upon happenings during his tenure. Most regrettably, Bill Corcoran and Joe Martin passed away in 1982. We're fortunate to have had their contributions in hand.

Finally, we have reprinted the final Opinion and Comment from Secretary Steve Tyler, also deceased, upon his retirement in 1954, and the final News and Notes Secretary F.J. Van Antwerpen prepared when he retired in 1978, both having appeared in *Chemical Engineering Progress.*

Larry Resen, Editor

George E. Holbrook
—1958—

While contemplating my assignment to write a personal anecdote, I recognized my handicap of having little interest in converting historical events into the written literature, and my bias of interest in history, almost extensively, to using it as a "potpourri" from which we struggle to find ideas and inspiration to generate new ones which, when developed and implemented, will lead to a more rewarding future.

I had the great pleasure in earlier years as chairman of a committee on "The Future of the Institute," to do just that in collaboration with those wonderful friends and associates, namely, Chaplin Tyler, John Perry, and Kerr Kettenring. That effort, with all its faults and shortcomings, led me to the conclusion that the germination, development and realization of worthy achievements of the Institute would be primarily dependent upon its organizational structure and the quality of leadership provided by its "rolled over" officers and committees—but even more importantly by the leadership, innovative capacity, and dedication of the headquarters staff. This line of reasoning leads one inevitably to the focal point of this staff effort, namely, the Executive Secretary. For part of the period of time under consideration we have been blessed with the incumbency of Franklin J. Van Antwerpen. He had the wisdom to surround himself with capable, dedicated and wonderful people—unfortunately too numerous to be cited individually.

Van and his lovely wife, Dorothy, who gave so much voluntary and effective help, must comprise the centerpiece of the Institute history from the Golden Jubilee of 1958 to its Diamond Jubilee of 1983. To have been closely associated with them and the "office family" for practically all of the Institute's third quarter century will always remain as one of the most rewarding and cherished facets of my life.

If these "observations" survive the deserved editorial treatment may they at least clearly reveal both sincerity and the merit of brevity.

Donald L. Katz
—1959—

My three years on the Executive Committee started in 1958, the year of our 50th anniversary. Our Golden Jubilee in Philadelphia in June was very impressive with George Holbrook at the helm.

Much advanced planning was involved for the Golden Jubilee June 19–22, 1958, for both ceremonial events and a technical program which addressed the shape of the future. Recollection of my personal involvement brought forth the experience of visiting with Warren K. Lewis before his leadoff paper on the session, "Where are we going in Chemical Engineering Education". Then I had the privilege of following him with a paper on the development of our program at the University of Michigan.

Many national leaders spoke at luncheons and dinners, such as Lee DuBridge, Crawford Greenwalt and M.E. Spaght. At the Tuesday luncheon as vice-president, I presided and introduced A.B. Kinzel from our brother society, the AIME.

Council had created the Founders Award to be initiated on this occasion, electing T.H. Chilton, J.V.N. Dorr, O.A. Hougen, S.D. Kirpatrick and W.K. Lewis to receive it.

It might be said that the emphasis on looking forward at our Golden Jubilee brought forth the concept of the Dynamic Objectives for Chemical Engineering which Council initiated in 1959.

Coincident with our 50th anniversary, the Institute became the fifth Founder Society along with the ASCE, AIME, ASME, and the AIEE. This was a beginning of our joint efforts to plan and finance the United Engineering Center. Following President Herbert Hoover, I joined the presidents of the Founder Societies, turning a spade of earth at the ground breaking in 1959.

Recollections of the period of my activity with Council brought the inauguration of divisions in the Institute. The proliferation of new disciplines and growth of technical activities showed the need for groupings at the organizational level for areas already delineated in the program committee. The International Congress on Nuclear Engineering held at the University of Michigan in 1954 was developed by the newly formed

Nuclear Engineering Division. Ours was the first cosmopolitan meeting looking at peaceful uses of atomic energy by 1,076 persons with 117 from 20 other countries. Then, the need for a division to match ASME's efforts in the annual joint Heat Transfer meeting resulted in our second division prior to the Golden Jubilee—a forerunner of many more divisions. Another item which reminded me of our forthcoming involvement in the computer age was reference to sessions on information processing and establishment of a committee on machine computation.

The developments in the last 25 years in the AIChE were undergirded in many ways for the forward looking views of Council as reflected in the Golden Jubilee period.

Jerry McAfee
—1960—

As the Fabulous Fifties yielded to be Sensational Sixties, the AIChE had another great year in 1960. At least it was a great year—one of the very best!—for the fellow who was privileged to serve as President that year. Among other pleasures, the Louisville Section arranged for me to be appointed a Kentucky Colonel; and the Tulsa Section persuaded the Osage Indians to elect me an honorary member of the Tribe, complete with head-dress, which I use to this day when I carve the Thanksgiving turkey. Regrettably, no troops were assigned to my command, and my Osage membership papers specifically excluded mineral rights.

Some of the highlights of that year which have survived in the memory bank include:

- In Mexico City the Institute held its first international meeting, enhanced by a bull fight, a stirring display of Aztec dances, and a speech *en Esplanõl* (!) by the unilingual President. That meeting, together with Secretary Van Antwerpen's trip to Europe later in the year, began the Institute's involvement in international activities.
- The cornerstone for the United Engineering Center in New York was laid, and the leadership of the AIChE in bringing that project to successful fruition and its example in

being the first of the engineering societies to reach its fund-raising goal were acknowledged.
- The arrangements were finalized for the first AIChE Petrochemical and Refining Exposition, which was held in February, 1961 in New Orleans.
- The Committee on Dynamic Objectives really got rolling and provided the focus for an enormous amount of deep thought and constructive debate about the future of the chemical engineering profession and the Institute's proper role in that future.
- Efforts to enhance the professional stature and status of chemical engineering—begun years earlier and expanded subsequently—included wide distribution of the pamphlet *Professional Standards* and a series of meetings with industrial leaders in connection with National and Annual meetings. In these meetings it was stressed that professional development encompasses a man's attitude of mind as well as his ethics and that there are heavy responsibilities both on employers toward their professional employees and on professional employees toward their respective employers.
- The AIChE formally entered the Computer Age through the activities of the Machine Computation Committee.
- It was clearly recognized that the future of the profession and many other values lie in encouraging the right kind of people to enter the profession and to establish and maintain an involvement with their professional society. To this end student chapter, local section, and career guidance activities were stepped up.
- The Institute enjoyed the largest membership gain its history to that date through the dedicated efforts of the Membership Committee and many others.
- Steps were taken to improve the AIChE's financial condition by encouraging endowment bequests and voluntary contributions in addition to dues.
- Responding to a recommendation of a special committee on services to members, the practice was begun of sending a letter several times each year directly from the President of the Institute to each member covering especially important AIChE policies, decisions, and actions.
- Executive Secretary F. J. Van Antwerpen turned in another outstanding year as "Mr. Chemical Engineering" and solidified his position as the dean of engineering society executive secretaries. Under Van's leadership, assisted by Larry Resen, Joel Henry, Sylvia Fourdrinier, and others, the staff of the Institute proved itself once again to be one of the Institute's major assets.
- Council effectively continued its long-standing tradition of careful consideration and debate of policies and programs designed to improve the chemical engineering profession and the Institute.

A comprehensive report would have to include many other items, some perhaps at least as significant as those mentioned above. The over-riding characteristic of the AIChE in 1960 was that, in common with other years, it provided the best available opportunity for chemical engineers to work together to achieve desirable objectives which would have been impossible to realize individually.

John J. Healy
—1961—

If one were to categorize past Presidents of AIChE, John Healy would fall under that of "the happiest fellow." He enjoyed life. He didn't put much stock in pretentiousness, on or off the job. He and his wife Maisie had a bubbling quality which brightened up the room they might enter. He certainly brightened up AIChE Council meetings.

This facility did not prevent him from being most effective in his professional career, both industrially and with the Institute. A lot of things happened within the Institute during John's tenure as President, some having been spawned during his vice presidency. Perhaps the biggest achievement of the Institute was the launching of the first AIChE Petrochemical and Refining Exposition in New Orleans in 1961. As an aside, there were limited facilities in the Crescent City, and the AIChE Show moved in right after the Stock Show moved out. Some exhibitors occupied space which most previously had been used for cattle stalls. John said the fragrance would separate the sheep from the goats among the exhibitors and attendees.

Two other items of note: the first issue of *International Chemical Engineering* saw the light of day and the Dynamic Objectives Report came to fruition with its being published in *CEP*.

The Institute also began its publication and usage of the keyword-index and concept-coordinate method of information storage and retrieval. All articles had a list of the keywords coded with the ultimate thought being that by matching them with the proper concept, all knowledge would be conveniently available on the subject. John didn't hold much with the idea, he figured you could find what you needed other ways. Apparently most others felt that way for the system had a relatively short life.

One final remembrance. His company got on an economy kick and came out with an edict that there would be no more first class air travel permitted for anyone. John was a senior V.P. and used to his perks. He said that no matter, he would pay the difference. By golly, he was going to go first class. And that was John—first class all the way.

Larry Resen

John J. McKetta
—1962—

The whole year of 1962 was one great big highlight for this ex-coal miner and farmer turned college professor. It was a year of great pride and personal satisfaction. First, I was proud to have been elected to this highest position in my profession. Second, it was a real pleasure to work with such a dedicated staff as Van Antwerpen, Larry Resen, Joel Henry, Sylvia Fourdrinier, Valeria Pastyik, and many, many others.

I had the pleasure of visiting every section that year with the encouragement of both Van Antwerpen and Larry Resen.

A disappointment that I had turned out to be a great pleasure. It seems as though that Van arranged to take the current President on one international trip each year. This year he refused to go with McKetta so he sent Larry Resen with me to Japan. There was some superficial purpose such as starting an *International Chemical Engineering Journal,* but we heard later that the real purpose was that they were hoping that McKetta and Resen would be lost forever in one of the baths (which we frequented often) or somewhere enroute. At the early stage of this visit Larry Resen hurt his back and I made the mistake of carrying his bag. As a result, everywhere we went, the Japanese people automatically assumed that he was the senior officer and I was his slave. I never caught onto this and Larry, who did catch on, never did tell the people any different. In fact, even when I tipped the bell boy (or bell girl whatever the case may be) they would turn toward him and bow and thank him for the tip.

There were no major achievements during my year as President. The Council felt that we desperately needed an increase in dues and that we would ask for one at the end of the year. It was for this reason that Van and I agreed that the President should visit each section to try to make sure the morale and esprit de corps were there so that the new increase in dues would not be with a great drop in our membership.

I have always suspected but I never really knew until my Presidency of the great devotion of members of the permanent staff of AIChE to our society and to our profession. Even though they were underpaid, and had to live in New York City (which is up north) I was tremendously impressed with their high esprit de corps.

I wish all of my AIChE colleagues could see the picture of Larry Resen that I took in one of the baths in Tokyo.

W. R. Marshall
—1963—

The year was 1961. The country was Japan. The time and place were November at the Osaka Hotel. The occasion was the 25th anniversary of the Society of Chemical Engineers, Japan. A special event occurred during hotel check-in. It was the receipt of a cablegram from F. J. Van Antwerpen which informed me I had been elected Vice President of AIChE for 1962. What great news! Twice before, my attempts to be elected had been thwarted. First by Jerry McAfee, and then by Johnny McKetta. The third try was a home run all the way to Osaka. After reading the cablegram, I wondered if Van Antwerpen, with whom I was destined to work closely for 15 years, really knew this person Marshall who had just been elected? Could he get along with him? Later I was to learn that he nearly always had misgivings about every new Vice President. And always he learned to understand their strengths, weaknesses, prejudices and especially their sports preferences.

My entrance as President in 1963 was preceded by the first AIChE presidential retreat with the Executive Secretary. It was at Syracuse in the dead of winter! This retreat was the first time that the President-elect and the Institute's secretary met for two days to review the problems of AIChE, to plan the forthcoming year, and to get to know each other. It was most productive, and this activity has continued to the present. One outcome of this retreat was a plan to put all of AIChE records on microfilm for preservation and for posterity.

The year 1963 was memorable for many reasons. I found myself conducting my first Council meeting in New Orleans with a miserable cold. It was a two-day meeting and probably the last one of its kind. I did not realize that Executive Committee minutes were essentially automatic consent and so I dragged Council, for two hours, through every item!

1963 had a twofold challenge. I was President of AIChE and President of Associated Midwest Universities through June 30th. It became convenient to schedule meetings of both organizations in the O'Hare area. The Flying Carpet Motel became a frequent meeting place for Council, and the Executive Committee.

In 1962, McKetta endeavored to visit every local section. It did not seem to be feasible or useful to repeat this tour de force. So, I chose the device of bringing local section officers together in one place. Three meetings were held, two in New York, at the United Engineering Center, and one at Salt Lake City. I have been especially pleased that the Local Section Officers Meeting has continued strong 20 years later. It has cemented a strong relationship between the National Office and the local sections. (*This program was expanded to include divisions and national committees. Ed.*)

Probably the most satisfying recollection of 1963 was the inauguration of AIChE's Continuing Education program. The idea that engineers need continued opportunities to learn about new chemical engineering theory, methods, and practice became very obvious during this year.

The efforts to persuade Council that this was true became a real challenge. Council considered whether AIChE should be involved in continuing education at nearly every meeting. It always seemed that more study was needed. Council could not be overpowered! It has its own perceptions, judgment, wisdom, and inertia. I attest to Council's wisdom however, because at Houston, November, 1963, it passed a motion to establish a Continuing Education Program. In 1964, President Don Dahlstrom appointed a committee to implement this program. The Institute through this committee has subsequently developed *the* outstanding continuing education program among professional societies.

The conclusion of my year as President was especially satisfying and rewarding. In December of 1963, Joel Henry and I traveled to London to plan a joint meeting in 1965 with the Institution of Chemical Engineers. It was extremely challenging to meld two systems for holding meetings. The British had misgivings about holding simultaneous sessions, and they were very worried about meeting in the newly constructed Hilton because it was so tall some occupants could look directly down into Buckingham Palace and the Queen's private rooms. The 1965 joint meeting was the first of many joint foreign meetings which were destined to occur, at least one in the term of every succeeding President.

Donald A. Dahlstrom
—1964—

As years pass, memories fade but I have done my best to recall major achievements that occurred in the years 1963-65 when I was on the Executive Committee. During these years, as well as all of the other years I have had the privilege of being a member, there was outstanding progress. In no particular order, I can remember the following:

Four divisions of the Institute experienced their birth pains and eventually became full divisions. At that time we only had two divisions, indicating the rapid growth.

Continuing Education started under the very capable leadership of W. R. "Bob" Marshall. Its growth to the fantastic program we have today for our members is obvious.

Our international effort began then and resulted in the Interamerican Confederation of Chemical Engineering, the Pacific Chemical Engineering Congress (PACHEC) and the many activities we now have with European chemical engineering societies.

The realization of the necessity of our interacting with the Federal Government resulted in the formation of the Government Programs Advisory Committee (GPAC). Obviously, the importance of this decision is very apparent today.

The Industrial Associates Program was started to raise money for needed research. The first effort was "Estimation of Physical Properties." We also realized that we could not do research of a proper quality with volunteers. A contract was made with Arthur D. Little and a recommendation to future boards to carry on in the same manner.

During this period, a decreased enrollment in chemical engineering at the university level was experienced and at the same time we experienced a severe demand. Money was raised from industry and individuals to produce the AIChE film to show the potential students the exciting challenge and opportunity in chemical engineering. We had a gala premier at the Pittsburgh meeting and obtained a great deal of interest through individual member activities in the high school as well as television.

To encourage membership growth, we lowered the entrance fees for new members and somewhat later they were entirely eliminated. It is apparent we continue to see membership growth adding continually to our strength as well as services rendered to members.

While many humorous things occurred during these years, the Puerto Rico meeting stands out in my mind. We had an important man from another country as a part of our program and it was my assignment to introduce him at one of the luncheons. I also introduced his wife in the audience who turned out to be his secretary instead.

Also, Bob Marshall and I and our wives took a tour around Puerto Rico and picked up a pastilla from one of the roadside stands. We decided that F. J. Van Antwerpen should be the recipient and left it in his hotel box. We waited all during the dinner for him to comment on this "delicate" item to get his reaction. However, he won the test. When it came time for dessert, the waiter brought the pastilla and Van proceeded to put ketchup on it and devour it.

In closing, I am certainly extremely appreciative of the opportunity I had to work with so many outstanding men during this period. Such compensation could never be exceeded.

W. B. Franklin
—1965—

Major emphasis during my tenure on the Executive Committee was on relations with local sections, international activities, and the establishment of more divisions.

Problems had been developing in relations between local sections and national for a number of years. During his presidency Johnny McKetta did a monumental job in visiting every local section then in existence, and did much to improve relations. At that time there was no requirement that local section members be members of AIChE and the sections behaved much like independent clubs. After lenghthy negotiations between Council and Local Section leaders, it was agreed that all Local Section charters contain a clause requiring membership in AIChE. Council then agreed to a plan of financial assistance to Local Sections to make it possible for them to communicate Local Section activities and meeting plans with every member in their area. The rebates were especially helpful in strengthening the programs of the small sections. These changes along with annual meetings of Local Section officers with National officers and the AIChE staff for one or two days greatly improved communications and cooperation between the sections and national.

Several years prior to my term Council decided that AIChE should take the lead in encouraging close relations with chemical engineers in other countries. This directive resulted in plans for joint meetings with societies of other countries. A most important Joint Meeting between AIChE and the Institution of Chemical Engineers was held in London in June of 1965. It was a huge success with fine technical sessions, a joint meeting of Councils of the two societies, formal banquet and a great social program so I am told. Unfortunately, I suffered a heart attack in April and did not recover in time to attend. Vice-President Stu Churchill presided in my place.

Later in the year I represented AIChE at a World Conference on Chemical Engineering sponsored by the Mexican Institute of Chemical Engineers in Mexico City. Secretary Van Antwerpen and I represented Institute at the European and United States Engineering Conference at Stockholm, Sweden. Plans were initiated for another World Conference; a tripartite meeting of the Canadians, the British and the AIChE; and a second meeting with the Mexican Institute of Chemical Engineers. An outgrowth of the international activity were plans for reciprocal membership privileges for visiting chemical engineers and reciprocally reduced subscription rates for publications.

Council recognized that the Institute had grown to the point where it could support additional Divisions. Certain members of Council were appointed to determine the interest in several potential areas. As a result Hugh Guthrie was successful in establishing the Petroleum and Petrochemical Division and Past President Don Dahlstrom the Food and Bioengineering Division before the end of the year.

Some important new first steps were taken by Council in the area of government relations. It agreed in principle to have an AIChE representative appear before a Government body—in the first case on a patent question. This opened the way for chemical engineers to speak to Government on technical problems. Also a new permanent committee was established named the Government Programs Advisory Committee.

It wasn't all work. The association with the cream of the profession, with Van and the excellent staff of the Institute was rewarding and inspiring. Most fun were the Executive Committee meetings on George Holbrook's sailboat on Chesapeake Bay.

SEVENTY-FIVE YEARS OF PROGRESS

Stuart W. Churchill
—1966—

One of the more enjoyable functions of the officers of the AIChE is to visit with the Local Sections. During my tenure in 1965-67, this was a particularly important activity because we were trying to convince the Local Sections to require that their members join the AIChE. In return, we proposed some relaxation in the requirements for AIChE membership, and a dues rebate to compensate for their possible loss in local dues.

Murphy's law was alive and well on some of those visits. Van used to send us slides and prepared talks on AIChE matters. However, I spurned these in favor of technical talks which I could give on short notice and without notes. That occasionally got me in trouble when I was informed as I rose to speak that this was Family Night or All-Engineers Night.

I agreed on one occasion, with one-day notice, to substitute at the Permian Section meeting for then-President Bee (Dr. W. B.) Franklin who was ill. They insisted I give his talk. Van said: "No problem! Bee will send his slides to Odessa and you can review them on your arrival." I was doubtful, but in a weak moment gave in. Of course, I arrived only moments before the meeting and realized that I would have to "wing" the talk from the slides. Unfortunately, the projector jammed on the first slide and the bulb of the replacement projector burned out on the second. My memory is happily blank concerning the rest of that evening, but I do remember having to scramble across the country to give a talk at the Midland (Michigan) Section the next evening. Past-President Don Dahlstrom also attended, and I offered him a ride back to Ann Arbor after the meeting. We had a puncture and he graciously helped me change a tire in a cold downpour. We briefly discussed the risk of driving without a spare, but decided to forego a pit-stop for tire repair in view of the late hour and the low probability of a second puncture. Probability took a beating—fortunately within jogging distance of my home.

One February I arrived in Cincinnati in the midst of a heavy late-afternoon snowfall to find that the Ohio Valley Section had cancelled their meeting. Ray Katzen said if I could travel 300 miles in the snow they could manage a few blocks or miles, and somehow reactivated the meeting.

That same year I was the banquet speaker at the St. Louis Section All-Day Technical Meeting. After the technical sessions, complimentary martinis were poured like water from pitchers. As the dinner began the power failed and we were informed it would not be back on for 45 minutes. The Section Chairman leaned over in the dark and suggested that we salvage that interval with my talk. Since I was planning to lean heavily on my slides, I said "No way." Again, I was persuaded against my better judgement. My opening remarks about the audience imagining the first slide in the dark were greeted with boisterous laughter. So was my second sentence. In fact, everything I said was accepted as side-splitting humor. Despite this success I was becoming a little desperate by the time the lights came back on. That speech would not have been such a success, even with the slides, had it not been for the generous infusion of martinis.

The opportunity to meet so many of my fellow chemical engineers was a great privilege and more than compensated for all of the travel and such nervous moments as mentioned above.

Theodore A. Burtis
—1967—

A backward look some 15 years after starting to serve on the Executive Committee produces for me more in the way of impressions than sharp images of events. We were worried about growth, although growing; concerned about communications, although communicating pretty well. In 1966 we had only one division and agonized over the potential "divisiveness" of divisions, a thought which I suspect now worries no one. Nevertheless, we felt very venturesome in granting status to the Food and Bioengineering and the Petroleum and Petrochemicals Divisions.

AIChE was the smallest, newest and by far the most independent of the Founder Societies. This did not always earn us the affection of our professional colleagues, as when we dropped out of the EJC rather than accept what we considered was a gross distortion of its character and purpose. We may not always have been right, but I believe that the willingness of AIChE to stand up for its principles was and is a source of strength.

Serving on the Executive Committee was a job—and a time consuming one at that. But it had its rewards. One was that the debates were a lot less tedious than some of those in Council. Even better were the fall sailing weekends on the Chesapeake on George Holbrook's "Wolverine".

Surely no one ever showed more devotion to AIChE than George did by crewing his beautiful yacht with left-footed landlubbers the likes of Churchill, Peters, Van Antwerpen and Burtis.

Max S. Peters
—1968—

During my term as President of AIChE in 1968 and my period on the Executive Committee, there were many exciting, serious, humorous, and important events in which I was involved. I could write many pages on these various occurrences, but I think it would be most appropriate for these anecdotal notes to comment on one or two of the humorous events that occurred during my term which helped to lighten up some of the more serious moments. Without doubt, the funniest event that ever occurred during my term on the Executive Committee was in the introduction of George Holbrook at one of our Board of Directors luncheons and the reference to his listing in *Who's Who in America*. However, I think the details of that event should be held for private telling, and I would like to tell about another humorous and somewhat embarassing event that occurred during one of our luncheons where I was the master of ceremonies.

The luncheon was held in 1967 at the National Meeting in Tampa, Fla., and it was the first regular luncheon where I served as master of ceremonies as Vice President of the Institute. For these luncheons, there are always a number of people at the head table whom the master of ceremonies introduces. In this particular case, there was a visiting dignitary from a chemical engineering group outside of the United States who was added to the head table at the last minute; therefore, his name was not on the list provided to me by Joel Henry, and I had written it in the margin of the sheet. In addition to the normal introductions of the people at the head table, I was supposed to read a special proclamation signed by the Governor of Florida which said something about chemical engineers being the greatest and dedicating the day to chemical engineering.

Prior to these luncheons there is always a cocktail period where I normally had one or two mild drinks. For some reason, on that day, I had two martinis and I paid the price. As I was going through the list of introductions for the people at the head table, I failed to notice the name of the foreign dignitary on the side of the page and Van had to rush up to the head table to remind me about introducing him. That was the only problem I had until we got to the place where I had finished reading the flowery proclamation sheet about how great chemical engineers were. After reading the proclamation, I proceeded to announce, "Signed by the Governor of that *Great State of Colorado.*" Needless to say, from then on, one martini was my absolute top limit.

Hugh D. Guthrie
—1969—

Three years on Council as a Director immediately preceded the three years on the Executive Committee in 1968-1970, and those six years blend together as one period. They represent a wealth of memories. One that stands out is the personal commitment and dedication characteristic of all Council members. Placing first priority on Council was the rule without exception. During those six years there was a total of two absences from Council meetings. Both were for illness. Al Smith making a round trip from Greece just for a Council meeting exemplified that attitude.

Those years are a blend of people and events. For me like pins on a war map were visits to local sections and student chapters during that period. I probably addressed more of them than any President except the inimitable John J. McKetta. There are always trying moments—as when the Muzak in the restaurant cannot be shut off, and the background music suddenly changes from the Blue Danube Waltz to The Yellow Rose of Texas. Of all the memories I probably value these most. Always impressive was the dedication of a small group of officers giving their responsibility to AIChE a first priority and making a local section or student chapter vital for its members. It was an opportunity to see where the dedication and commitment that characterizes Council begins. Starkly standing out is Bruce and Phyllis Martin taking me with them by car from Cincinnati to the Indianapolis meeting, and as a result my not being on that ill-fated TWA flight when all 88 aboard were killed.

Van and Dorothy—a team of key importance to every AIChE President. My wife, Betty, described the year as President as 100% for AIChE, 100% for Shell, and minus 100% for the family. Van and Dorothy did so much to make that year a survivable one for us. Always available, always helpful was Joel Henry—superbly organized, covering all the intricate details in schedules not having a minute's flexibility.

No reminiscence would be complete without recalling the impact of George E. Holbrook. Like E. F. Hutton—when George spoke, everybody listened. One of the still-remembered pieces of wisdom he gave me was a quote he attributed to Winston Churchill, "Conflict between societies, nations, groups or individuals rarely occurs over differences on *ends*. Nearly always the conflict stems from difference on *means*." He counseled that in many situations complete agreement on *ends* may overlook critical differences on *means*, and in those circumstances never to underestimate the influence of a quietly spoken but firm "No." There were several times when that advice was appropriate during the year as President. George also is remembered for the once each year Executive Committee meetings on his 44-foot yawl in Chesapeake Bay. Awakening with Messrs. Holbrook, Burtis, Peters, and Van Antwerpen still asleep was an experience I shall never forget. The combined snores from that august group can best be described as an out-of-tune engine operating under widely varying load.

Those were the years when Ted Burtis spearheaded the thrust to Divisions in AIChE and for each Division there were strong individuals making it a reality. In that group of key individuals Larry Cecil amazed all of us. His endurance and energy would have been difficult for a man 40 years his junior to match. No other event in those three years so strongly emphasized the impact one man can have when he is completely dedicated to a clearly defined goal.

The three years on the Executive Committee are part of a continuum of 35 years in AIChE. No letter can cover the many memories represented. I recall the line of one of the inmates in Tolstoy's "Asylum of the Night"—"we are all just human beings." That is true, but chemical engineers have an important distinction—they are chemical engineers and AIChE provides a framework for sharing that special difference. A quote I've often used is a remark of Frederic Dannerth. In his 90s and then the only living charter member of AIChE, he was asked what one sentence he would give as advice to a young chemical engineer. His response was, "I would tell him to remember that you never fill a job, you create one." That is true not only of one's career. It is also true of AIChE. Opportunities are what the individual makes of them.

Arthur L. Conn
—1970—

As I think back on the three years spent on the AIChE Executive Committee as Vice President, President, and Past President, I believe one of the most significant developments during that period was the increasing recognition of the chemical engineer's relationship to society as a whole. At that time, we added a third objective to Article II of the constitution—"to serve society, particularly where chemical engineering can contribute to the public interest." The Institute has followed through on this in many ways since. I chose it as my area of emphasis during my year as President—in communications with Council and with the entire membership, as well as in speeches to local sections. Some concrete accomplishments were the formation of the Environmental Division, with greater emphasis in programming on pollution control; greater participation of local sections in community projects; and increased efforts in career guidance to the disadvantaged. Following up a letter that Hugh Guthrie had sent to President Nixon, we proposed to the Office of Science and Technology several projects in which the AIChE was prepared to advise the government. This was followed up by later administrations.

In more recent years, many other steps have been taken to implement this objective—for example the many activities of the Government Programs Steering Committee, the establishment of an AIChE Washington office to maintain better contact with the government, the formation of committees to adopt policies on nuclear wastes and on energy, and sessions at annual meetings discussing how chemical engineers can participate more effectively in acquainting the public with controversial technical issues. So a concept that first resulted in a constitutional change has had a marked effect on the Institute's activities, as well as on some of my own. I look forward to the day when there will be at least one member of the AIChE in Congress, who will not only initiate important legislation, but also bring greater understanding of technical issues to the entire legislative body.

Some of the most enjoyable experiences during that three year period were the contacts with a fine group of people in Council and in the membership at large. While we worked hard at Council meetings, we had to handle some of the stickier problems at the Executive Committee meetings. But at these meetings we also enjoyed a camaraderie beyond our work. Among the memorable experiences were: sailing on George Holbrook's boat on beautiful Chesapeake Bay, watching Michigan's (then) invincible football team in Ann Arbor, and getting reacquainted with the lore of the Revolution in Williamsburg, Va., just to name a few. And I still remember a delightful weekend several years later, with a meeting of past presidents in the Ozarks to discuss dynamic objectives and more importantly to celebrate Van's 20 years as Executive Secretary. It has been exciting to have participated in the growth of the Institute over more than three decades. From a group interested mainly in unit operations and unit processes the Institute has become an organization that is also concerned with the personal development of the engineer all through his career, while reaching out to engineers the world over and actively contributing to society as a whole.

Joseph J. Martin
—1971—

When Editor Larry Resen requested that AIChE presidents during the past quarter century relate an experience during their tenure on the Executive Committee, it was only natural to look through old files to retrieve some information that hopefully might be of interest to someone with the courage and fortitude to plow through the 1983 Diamond Jubilee publication on the history of the Institute. What came from the files gave me the pleasure of recalling and re-living past memories, and I can only trust that you, the reader, may gain a fraction of that pleasure if you have the persistence to read through this small portion of history.

Every President likes to think that he left some indelible stamp on the Institute that continues long after he has bowed out of the mainstream of its activities. One particular incident that I recall is so insignificant that it hardly seems worth recounting. Most of you know that Council meets at every national meeting and spends at least a full day (normally the Saturday before the meeting starts) going through the action items that need attention by this policy-making body of the Institute. How many of you also know that Council at this writing has an informal session on the preceding Friday night where a give and take discussion is held on topics of interest that may be coming up in future AIChE operations? Well, the Friday night session was not always so, even though the formal Council meeting on Saturday was of longer standing than my memory could recollect. But, therein, lay a problem. During the four years on Council before becoming President, I found myself frustrated with the formal session, hemmed in by strict Council procedures and bowing before the ever-present god of Roberts' Rules of Order with detailed minutes recorded for the whole meeting and little chance to let down one's hair and philosophize off the record. Consequently, I suggested that during my tenure we should gather on Friday night and engage in what I termed a "bull session" wherein we would carry on free discussions of upcoming problems of the Institute with the simple restriction being that the chair would permit only one person at a time to air his pet thoughts on a subject. Coincidentally, Vice-President Tommy Tomkowit had been working with the Member Relations Committee and they had essentially the same suggestion; i.e., that Council should initiate formal "brainstorming" sessions on topics that were not part of the formal agenda of the Saturday Council meeting. With this fortuitous agreement in objective, it was not difficult to persuade Council members to come a few hours earlier on Friday evening for the discussion we contemplated. In all of this planning Tommy continued to refer to "brainstorming" sessions and I stuck to "bull sessions," but eventually he won out with his designation that was more worthy of application to the dignified Council of the prestigious American Institute of Chemical Engineers. However, one item of correspondence showed that Tommy did not want to characterize the Friday night affair as a "B.S." session, so he called it a "B.A." session. Do you know that to this day I am not sure just what he meant by his B.A. acronym, though there never was any difficulty in knowing that we both were talking about the same kind of meeting.

Just to give the flavor of what was in mind for the "B.S." session, the following quotation is taken from a letter I wrote to Council preceding one of the meetings:

"The general subject for our session will be 'Whither or Wither AIChE—Technical or Professional Society or Both.' The frame of reference for this discussion is our history. We were founded as a scientific or technical society with the objectives being the 'advancement of chemical engineering in theory and practice, and the maintenance of a high professional standard. . .' We recently enlarged upon this to include 'to serve society. . .' To carry out these objectives, we have sponsored hundreds of technical meetings, published a vast amount of literature, conducted education programs, accredited curricula, set up standards of professional conduct, presented awards, been active in career guidance, supported research, discussed legislation and government directives with appropriate officials, and have modified our organizational structure to include Divisions, Local Sections, and Committees."

"Knowingly or not, we also have instituted a number of activities on behalf of our members that represent a stretch of imagination of interpretation of our objectives. Consider, for example, our life insurance program, our participation in Pensions for Professional Employment. These actions reveal concern for our members even though our Constitution does not."

"Where should AIChE go tomorrow? Should we swing the pendulum back to being more of a technical society or should we expand our efforts on behalf of our members? Should we change our IRS category and go in for active lobbying for member benefits? Should we place ourselves more in the position of bargaining agent or at least counsellor for members in their negotiations with company management? Should we become a union? Should we develop controls on the number of chemical engineers in the nation? Should our membership requirements be modified? Should we abrogate responsibility for the economic well-being and professional climate of our members and let some other group take over that job? Should we try to maintain a middle-of-the-road position as a dual-purpose professional-technical society that has the additional objective of being concerned about the welfare of its members?"

"Answers to the above questions will indicate where AIChE should go in the next decade. As a result of our discussion, we might well undertake a new member opinion survey and determine what kind of society they desire. Following that an internal study of 'Dynamic Objectives for AIChE' may be in order."

Believe it or not, some of these ideas did come to pass. Also I am told that in New York headquarters today they refer to the informal sessions as B.S. sessions, but whether B.S. means bull session or brainstorming or something else is not clear, but the modern day informal Friday night sessions are for the same kind of off-the-record discussion as were initiated over a decade ago. Thus, it is rather reassuring to know that one small accomplishment of the past era is still with us today.

T. W. Tomkowit
—1972—

I suppose every Director and Past President, as he reflects on his term of office, concludes that his administration identified many new opportunities and resolved many troublesome problems.

I found that the experience was very rewarding—a training experience in interpersonal relationships—a rededication to serve both society and our profession—a lesson in humility. Words cannot describe the personal satisfaction that I have realized through this in-depth involvement with my fellow professionals and our profession.

The following summary highlights a few of the activities in which we were involved that were particularly satisfying. The period covered was 1968 to 1973.

- Constitution—Expanded Objectives—"To serve society particularly where chemical engineering can contribute to the public interest."
- Career guidance for the disadvantaged—the need was recognized, Council agreed that it was essential that AIChE make a societal contribution. A subcommittee was established.
- DIMP (Design Institute for MultiPhase Properties) We finally gave birth to this worthwhile program. The concept has been expanded to establish funding for other research programs.
- Employment Guidelines—We developed our own criteria and were instrumental in committing Engineers Joint Council to develop intersociety guidelines.
- Intersociety Committee of Society Presidents—An informal committee met regularly in Washington to respond to the unemployment problem. We agreed that the meetings were valuable: discussed the desirability of continuing the exchange; AIChE proposed that we establish a new umbrella organization. The suggestion was not implemented, but it is conceivable that it might have been the seed that matured into the American Association of Engineering Societies.
- We participated in the Institution of Chemical Engineering Jubilee meeting in London and PACHEC I in Kyoto.
- One of our Council members died during his tenure. We deeply regretted the loss of Cecil Chilton, a strong member of Council and a credit to our profession.

One anecdote does come to mind—it occurred during our European tour. The Vans accompanied us to Rome. Our stay was very pleasant but each evening our dinner was cancelled because of a strike. The following day the hotel activity was normal. On the third night and third strike, Van took things into his hands. He proceeded into the kitchen and began to prepare dinner for us and other hotel guests. It was an hilarious experience. The Italians just could not believe what they saw. We know that Van's action reflects the expected resourcefulness of the chemical engineers.

Ann and I would like to extend our congratulations to all the members of the AIChE on this our Diamond Jubilee. I encourage the administration and members to seek new opportunities for the AIChE to serve our members, society, and profession.

SEVENTY-FIVE YEARS OF PROGRESS

Theodore Weaver
—1973—

Thinking back over the three years 1972 through 1974, I recall most strongly four major themes that occupied the Executive Committee and Council.

- *Informing Public Leaders and Opinion Makers of the Real Nature of Our National Energy Problem.* This took many forms and involved a number of Committees, but the images that still stick in my mind came from the series of discussion luncheons and dinners we hosted in major cities around the country. With 30 to 50 local political and business figures as our guests, I spoke on "Energy—The Largest Problem," to help them understand that we were entering a new era in which energy costs would be "high," and supplies less plentiful—that the causes were fundamental, not conspiratorial—and that we would all need to accept difficult choices to solve that problem. Then we invited discussion. In New Orleans a young man who had not used a comb in a long time got up and angrily denounced the hot combs which had recently become a popular fad. In Detroit an apparently successful businessman insisted that the problem was caused by our country's massive exports of petroleum, and refused to accept data to the contrary. In Chicago, a woman who was trying to save Lake Michigan, earnestly explained that at times she had found it necessary to adopt unreasonable, short-term positions in order to achieve reasonable, long-term goals. And in Indianapolis a General Motors executive told me that a target date had already been set (in 1972!) to end production of GM's largest automobile bodies—for Cadillacs, Oldsmobiles, and Buicks.
- *The Obvious Need of Governmental Bodies at All Levels for Sound Technical Information, Effectively Presented.* In 1974 I served as chairman of a committee of dedicated members that studied a number of approaches to the problem of interacting with the Federal Government. As a result of that effort, a full-time Washington presence was established, under the guidance of a new Government Programs Steering Committee, to find opportunities for helping our Divisions and Committees to deliver sound technical information in a useful way. It was also concluded that we would *not* make a regular practice of supporting or opposing specific legislation.

- *The Expanding Domain of Chemical Engineering.* Our members were becoming increasingly involved with applying chemical engineering to entirely new activities—bioengineering, environmental protection, synthetic fuels—and there were forecasts of biosynthetic process engineering ahead. This "expanding domain" was one of the factors that led to the long-range planning effort that became the Second Dynamic Objectives Study. It also stimulated the Program Committee to describe our National Meetings as opportunities for A MASSIVE LEARNING EXPERIENCE.
- *AIChE's Proper Role in Employer/Employee Relations.* During that time of significant unemployment and business stagnation, this issue was surely the most emotional (and potentially the most divisive) presented to Council. The great challenge was to channel the energy behind demands that the Institute "do something," into actions which, by consensus, would effectively serve the interests of substantially all the members. I believe that challenge still exists.

Irving Leibson
—1974—

The Institute's year of 1974 was characterized by significant emphasis on planning for the future. As the AIChE Vice President in 1973, I was charged with the responsibility to develop a forecast of the climate expected to prevail during the next five years. A Delphi analysis was conducted as a technique for exploring expert inputs via successive questionnaires to evaluate the future business, government, and professional climate for chemical engineering in order to determine the real needs of the practicing chemical engineer through the next decade.

The following objectives were established for the Institute, arising from a combination of the results of the Delphi analysis (which were published in the March, 1974, *Chemical Engineering Progress*) and the Friday night "Think" sessions with Council prior to each national meeting:

1. Attract qualified non-member chemical engineers to Institute membership.
2. Increase membership involvement in Institute activities at all levels of the organization.

3. Bring the opportunities and challenges of a chemical engineering career to the attention of young people through effective career guidance activities.
4. Develop a continuing mechanism for the preparation of realistic forecasts of future needs for chemical engineers.
5. Develop effective means for interacting with the Federal government to:
   a. make chemical engineering expertise available where it can contribute to the public interest.
   b. review and advise *re* proposed legislation and administrative rules.
   c. promote and support programs to enhance the engineering profession.
6. Develop new programs and continue current activities to achieve implementation of the Unity Guidelines to Professional Employment in the membership personal needs area.
7. Develop an effective Institute public relations activity to improve internal communications and to publicize the contributions of chemical engineers to societal problems.
8. Seek opportunities for cooperative and unified action with other professional and technical societies with the long term goal of unification of the engineering profession.
9. Review and evaluate all of the technical activities of the Institute including publications, meeting content and format, and other sponsored activities (e.g., continuing education) for relevance to the membership, cost-effectiveness, and value to the profession.
10. Continue the program of public education concerning the need for a national energy action program.
11. Initiate a study to forecast the technical skills that the future chemical engineer will require and provide for translation of the results into academic curriculum via accreditation programs.

The year ended with a decision by the Council that it would be timely to update the Dynamic Objectives Report for Chemical Engineering, and the incoming Vice President, K.D. Timmerhaus, was appointed to spearhead the organization of a new Dynamic Objectives Committee in order to carry out this mission.

The objectives listed above still seem pertinent to ongoing AIChE activities in the years to come.

SEVENTY-FIVE YEARS OF PROGRESS

Kenneth E. Coulter
—1975—

We had a few experiences which were amusing as we look back. They are as follows:

1. At the Munich Meeting Joel Henry was anxious to get the bus to the airport started early because of a big celebration in Munich. The Coulters were the last ones to get on the bus and when the bus got on the way to the airport the streets were almost empty.
2. When we got into New York we hit a traffic jam at the airport and missed our plane but caught another an hour or so later. On the trip to Detroit the pilot was too close and hit a spot where a plane had just crossed and he almost rolled it over. I am sure the passengers were ready to sing "Nearer My God to Thee."
3. At the World Federation of Engineering Organization in Tunis, Tunisia, we had some laughs. Van and Dorothy arrived at the Hotel Africa (the meeting place) a week ahead of us and when we checked in and asked for Van, we found they had gone to another hotel. When we got up to the room, Eleanor decided she wasn't going to stay but she finally did. Later, we found the door didn't lock and a few other things. Eleanor was afraid of the food and only ate omelets at the hotel.

    We were royally entertained all week. At a dinner given by the Minister of Industry we were at a table with either an Indian or Pakistani and when the dinner came out he said, "I want steak." Van and Dorothy tried to convince him that the food agreed with his religion, but he insisted and they gave him steak.

    At another gathering, they had an outdoor meeting complete with a native band and high-class belly dancers.

    One noon we went out to Carthage for lunch with the President of Tunisia and saw how he lived and enjoyed a many-course lunch.
4. In Caracas, Venezuela, we arrived one day early to find that the program was just being printed. Van and I had a meeting scheduled for 10 am Monday that wasn't held until Wednesday. I was to give a paper and after several days' delay, had 20 minutes' notice to give it. We sure found out that South American customs were different from

ours. The weather, food and accommodations were excellent.
5. We learned what the Third World Countries thought of the Industrialized Countries. They felt we sold obsolete technology, charged instead of giving. In North America, at the World Federation of Engineering Organization meeting, the Third World Countries voted with Russia, etc., when the U.S., Canada, France, England, Sweden, etc., were paying the bills. They tried to make things political instead of educational.

Now for AIChE accomplishments:

1. I was President when there was unrest among engineers, particularly in the East, so I decided we should talk to Management. We talked to the Technical Association of Pulp and Paper Institute (TAPPI), The American Petroleum Institute (API), twice to the Chemical Industry Board, and at a number of convention meetings where local executives were present. Sam West said at one introduction that I took the concerns of the engineers into the Board rooms.
2. At my last meeting, I suggested we start a Safety and Health Division, and appointed Gene DeHaven to guide it and it finally resulted in a division.
3. Helped set up initial meetings to get the Management Division started.
4. Asked the Universities (through the N.Y. Office) to list their areas of expertise and it was published.
5. In cooperation with the Mid-Michigan Section and a Government grant, a survey was made in the Midland area to see the rating practicing engineers gave their college courses.
6. Visited 85 cities on AIChE affairs.

Last but not least, I remember the great cooperation I received from Larry Resen, Van, Val Pastyik, Sylvia Fourdrinier, Joel Henry, and the other staff members. Joel was a great help to Eleanor and me especially in the entertainment area.

Klaus D. Timmerhaus
—1976—

1976 was a year in which the U.S. celebrated its bicentennial. It gave many Americans an opportunity to look back over our nation's illustrious, but stormy, history. It also prompted leaders in government and industry to review more recent trends and consider future directions of action. Such a reassessment was also underway in AIChE with the development of a set of objectives to guide the profession and the Institute during the following decade. This self-study entitled "Dynamic Objectives for Chemical Engineers" was launched in mid-1975 when eighteen Past-Presidents of the Institute gathered for a memorable weekend at a Lake of the Ozarks resort to honor F.J. Van Antwerpen for two decades of exemplary service to AIChE and to philosophically discuss the future of the Institute. The study, involving at times several hundred dedicated AIChE members and chaired by the 1976 Institute President, was completed in late 1976 and accepted early in the following year.

Those familiar with AIChE activities have recognized that some of the objectives developed in this second major self-study (the first occurred in 1959-61) were extensions of objectives that AIChE had been pursuing for many years, but which needed greater emphasis in the future. On the other hand, some of the objectives pointed AIChE in entirely new directions which could, in time, possibly alter the traditional image of the Institute. It goes without saying that these particular objectives were debated at considerable length and comments were invited from the entire membership before the final wording was accepted by Council. Upon acceptance, Council appointed a task force to develop plans for working toward these objectives, including identification of the committee or division that was to be charged with implementing such actions. Since these objectives engendered continuing activities by such entities, the task force was charged to develop realistic goals for each of the ten general objectives to permit a measure of progress in its implementation.

The bicentennial year also spurred the Institute's involvement with government and society. This greater involvement was highlighted by a reorganization of the Government

Programs Steering Committee and the formal establishment of a Washington, D.C. office. The objectives of the Committee, besides providing guidance to the Washington representative, were to develop effective mechanisms for providing the government with chemical engineering information, encourage AIChE members who were involved in assisting local and regional governmental entities with such information, and conduct an active communications program to demonstrate the Institute's concern and involvement in public affairs. One result of these objectives has been the establishment of some 60 local Government Interaction Committees. In addition to these activities, AIChE redirected its Speaker's Bureau to include more presentations on national issues. Similar redirections were made in the programming efforts of national and local meetings. Concern for public affairs was also evident in the deliberations of Council when they considered and debated such topics as the nuclear initiative, the Clean Air Act Amendment, the Offshore Technology Act, etc. Position papers were prepared on those topics where chemical engineering input was deemed to be of informational value to elected officials and their constituents.

In a similar vein, the reevaluation of the goals of the Institute also brought a greater concern for member well-being. The extensive layoffs of engineers, including chemical engineers, in the early 1970's brought great pressures on professional societies to serve in an adversary role relative to the economic welfare of their members. AIChE did not take such a step, since it realized that this action could destroy its creditability with industry and greatly weaken the purposes of the society. However, AIChE was not insensitive to the changing needs of its members and did join 27 other professional societies in setting-up reasonable guidelines for the employment of engineers. Since the development of these guidelines, AIChE has steadfastly maintained an advocacy role rather than an adversary one. However, continued pressure by members prompted AIChE to conduct an intensive survey of its members by an independent consulting firm to determine the attitudes of the members towards the guidelines and their desires for institute actions. This survey reaffirmed the advocacy position, but it also emphasized the desire of the members for greater Institute involvement in the implementation of these guidelines. As a consequence, AIChE launched an extensive program of not only educating its own members about the guidelines, but also scheduled a large number of regional meetings with industrial executives to express member's concerns and argue for their acceptance by industry. These discussions were followed-up by a Presidential letter to all industrial executives associated with companies employing 10 or more AIChE members. Positive responses were obtained from quite a few of these contacts.

Greatly increased freshman enrollments in chemical engineering of 43.4% in just one year also brought the repeated mismatches between supply and demand of engineers to the forefront in 1976. Since these cyclical mismatches are not healthy for the profession or the industry which employs chemical engineers, the Institute alerted all chemical engineering departments of the potential problem and initiated discussions to explore ways to ameliorate the impact of a possible oversupply of chemical engineers in the early 1980's.

The self-study also emphasized that if the Institute was to expand its activities, meet increased member needs, and improve the image of the profession, it would require more member involvement. Consequently, a revitalization of committees and divisions was undertaken. One phase of this program was to request committee and divisions to publish goals and objectives at the beginning of the year, with an audit by Council at the end of the year to determine whether these goals had been met. The study also showed areas that were not represented in the Institute. This led to the appointment of formation committees for two new divisions. To establish greater contact with local section members, the Local Section's Activities meeting was reestablished. Council minutes were sent to each local section, and 44 local sections were visited to encourage more member participation. In addition, two Presidential Newsletters were sent to all local and national officers to provide an awareness of the activities of the Institute.

Amidst all of these serious activities during the nation's bicentennial year and AIChE's 68th year there were many lighter moments that will linger in our memories for a long time. For example, those attending the Presidential Reception at the Kansas City AIChE Meeting may recall seeing various portions of the self-sticking wallpaper in the room begin to separate from the wall. When this was spotted, one of the taller attendees was requested to see if he could reaffix the wallpaper. It was a humorous sight to see him holding the wallpaper with one hand, while holding a drink in the other. However, as more and more sections of the wallpaper began to part company with the walls, more and more attendees were requested to locate themselves next to the walls until soon no one was in the middle of the room and all were next to the walls trying to keep the rest of the wallpaper from coming down. This was probably the first Presidential Reception that turned out to be a wallpaper hanging affair!

There is no question that the greatest pleasure that an AIChE President receives during his year of leading the Institute comes from the interaction that he has with AIChE members from all parts of the country. Each interaction, irrespective of the situation, makes him a little bit more proud of his professional society and yet more humble when he realizes all that has been accomplished by visionary leaders and dedicated members who have preceded him.

SEVENTY-FIVE YEARS OF PROGRESS

A. S. West
—1977—

Activities such as publications, meetings, continuing education, accreditation, and local sections continued to dominate the work of the Institute. These activities are primarily directed towards developing the technical excellence of the members. However, it became apparent that social, economic, and political forces—some controllable and some not— were starting to change the role of engineering societies. In 1977, the manifestation of this included a significantly increased interaction of AIChE with government entities and a substantially enlarged cooperative effort among the several major engineering societies.

These were not independent developments. Individual societies, with AIChE at the forefront, were increasingly involved in providing to the government appropriate engineering information and professional judgment on technical matters which affect public policies. Joint efforts were seen as a mechanism to strengthen the impact. The results of such cooperation are not only in the public interest but enhance and enrich the engineering profession as well.

The Presidents of the five Founder Societies—Scothe Kezios (ASME), Art Nedom (AIME), Bob Saunders (IEEE), Lee Walker (ASCE) and I—found a unique compatibility of interests and objectives. During several discussions, the concept for a new federation of engineering societies was generated. The five of us held a "secret" meeting in Houston to address this issue. Of course the meeting wasn't really secret nor was it meant to be, but it did provide the opportunity to discuss ideas freely and to get our thoughts on paper completely unencumbered with other subjects or influences. The precepts, together with the rough outline for operating bylaws, developed at this meeting led to the formation of the American Association of Engineering Societies (AAES) just two years later. Its organization could be considered an amalgamation of three existing federations with a sufficiently broader scope to attract all engineering societies. Perhaps at last, in the modern era, engineers can speak with a strong voice in the public affairs domain.

In another major intersociety event, Executive Director Van Antwerpen and I were part of a small U.S. delegation to attend the Congress of the World Federation of Engineering Organizations (WFEO) in Warsaw. Here, we managed a minor coup (for the times) in preventing the official seating of a delegation from Red China to replace the Taiwan engineers. The current situation regarding the Chinese delegations is significantly different today. The Peking delegation was seated in 1981 at Buenos Aires as the official Chinese voice, with Taiwan maintaining full member status.

The babble of tongues at the Congress was fascinating. The official languages were French and English. But imagine, for example, a Russian delegate—who could, incidentally, speak better English than he would allow—addressing the group in French with a Polish interpreter translating the talk into English. French with a Russian accent translated into English with a Polish accent does not make for readily understandable listening.

Reflections on 1977 would not be complete without indicating the unique privilege I had in working with both the Executive Director of many years, F. J. Van Antwerpen, and also with Chuck Forman who joined the Institute staff during the year to become the new Executive Director upon Van's retirement early in 1978.

William H. Corcoran
—1978—

For my three-year tour as a member of the Executive Committee, my fondest memory which continually brings a smile to my mind relates to the super qualities of Franklin J. Van Antwerpen. We take many things for granted as we move through our daily lives in meeting our responsibilities. One such item that especially is taken for granted is the need for the Executive Director of any organization to listen to the same material year-by-year as new officers move through the system and bring in their own pride and prejudices to deal with the same old problems in new and exciting ways. The big problem in dealing with these problems is that these officers believe they are creatively dealing with exciting and unique new issues whereas they actually are repeating, in most cases, studies and actions that have surfaced continually over the decades. There really is a uniqueness, but it is not found in the problem but rather the new time frame for the old problem. The patience of an Executive Director has to be infinite in such a situation because of the great press to reinvent the wheel. In such striving on our part, I paid particular attenton to the activities of Van in the course of my three years of work on the Executive Committee.

As we worked to do great new things, I suspect that I, in an unconscious effort to repeat the past, brought in old ideas well beyond any allowable quota. Van always treated my input with gentleness, patience, and thoughtfulness as if I had really invented something new. He was equally kind to all others. That receptiveness on his part made service on the Executive Committee a pleasure, but that fond remembrance is not the prime purpose of this brief essay. The prime purpose is to record for posterity my continuing amazement during the course of my work on the Executive Committee and my continuing amazement today at the fantastic ability he had to accomplish what he wanted to do in the presence of those gracefully received and great new ideas presented in avalanches by the new faces each new year. He never failed. His techniques were impeccable and 100-percent workable. I personally believe that Caesar, Charlemagne, von Clausewitz, Napoleon, Grant, Pershing, MacArthur, and Eisenhower would all have learned by study at his feet. All of these great leaders were successful because they not only knew what to do on the offense, but they understood defense. They learned how to appear another day to successfully carry out their campaigns. Van could have taught each one of them something new. Whenever an issue would appear before the Executive Committee that he thought should be decided contrary to the general thinking of the Executive Committee, he would hold his peace and keep his pleasant smile. No one can defeat a smile, and in my opinion no one ever defeated his. He would listen to all the same arguments that he had probably been listening to for the past 25 years about why something should be done, whereas in reality its successful implementation would be damaging to AIChE. He encircled such ideas and gently propelled them with his own plans as to what should happen.

Now my thoughts make members of the past Executive Committees and myself sound stupid. I really do not think we were. We had much to learn. Van let us learn, but he did not let us destroy ourselves. He used all sorts of fair tactics to make his point, and I think that the fairest tactic and the most powerful tactic he used would be to agree with what you said and then do what he had on his mind. That was fair because he was a loving, protective friend. I never got mad at him because he was always right. In the course of such exercises we learned much about patience and skill in moving in the direction of everyone's best interests. I will be forever indebted to him for this bit of useful style and knowledge that was added to my life. He did that for 25 years with all of the Executive Committee members and actually with all of the Council members.

So as I think of the great Generals in history, I would list them with Van in the lead and all the others that you can think of as second best. He truly has to be cited as being in the Vanguard in the useful development of ideas. We owe him much for his use of leadership and his teaching of patience.

James Y. Oldshue
—1979—

The theme that I feel best represents Chemical Engineering as we approach the Diamond Jubilee is that we chemical engineers need to participate actively in two worlds. I think that we are well founded in our participation in and understanding of the first world, the world of Nature, in which our background in chemistry and engineering makes us particularly valuable in this "chemical world." I may perhaps have emphasized the role of the chemical engineer a little too strongly, but I will defend this view on the basis of a little exaggeration which is close to the actual situation.

In the second world, the world of Society, I don't think we as chemical engineers in the aggregate play a large enough role. For one thing, there appears to many of us to be more than enough people manning the ramparts, shouting out and ensconcing all kinds of causes and fears. Environmentalists, human rights advocates, people concerned with social programs, equal rights for women and minorities, and many other causes seem to abound. Sometimes, these issues are very technical, and sometimes very obscure.

I would like to remind us that familiarity sometimes breeds contempt. Our technique of drawing flow diagrams for energy, material, and money around all of our problems to be solved is quite unique, even among many of our technical and scientific associates.

From my experience, this way of thinking is extremely valuable in keeping people who are concerned with government, social, and charitable careers and the general public aware of the projections and numbers of costs and raw material needs of planned programs. Our analysis will not always be welcomed on the committees, boards and conferences where the social fabric of America and the world is being debated, but I can assure you it is needed. It is not retrogressive, because an exciting but naive plan begun with erroneous bases can bring disaster to the very causes it was supposed to espouse.

We also need to keep striving to put economics, social security, and safety and health risks in perspective. A "no risk society" may be taking the greatest risk of all, in that we may end up with no viable society *remaining*. There are many ways of presenting risk data, and we need to think of all kinds of new techniques and methods of putting these things in perspective for the general public.

Hopefully, we also have a genuine desire to share our information, resources, and ideas with our fellow chemical engineers all over the world. Hopefully, this will be an altruistic sharing, with no serious worry about tit-for-tat return.

Our meeting with Mexico in 1980, and our meeting with China in 1982, plus our interest in setting up methods of exchanging technical information with all countries will be just a start with these mutually beneficial relationships.

Being a "two meter" chemical engineer who travels around the world in chemical engineering activities has also been a source of inspiration and pleasure.

James G. Knudsen
—1980—

Being an Officer of AIChE gives one the opportunity to observe at first hand the dedication of the large number of members whose activities really make the Institute what it is, a dynamic society striving to accomplish its objectives of serving both society and the chemical engineering profession. In addition, one can also observe and appreciate the loyalty and hard work of the staff in the headquarters office.

The Institute and the chemical engineering profession prospered during the period 1979-81. In 1979 emphasis was on the implementation of the Dynamic Objectives. An implementation plan involving all segments of the Institute had been developed in 1978. This provided an opportunity for an intensive self-study by the committees, divisions and local sections. Not surprisingly, many of these groups found that they were already implementing many of the Dynamic Objectives. In 1980, the Institute established a Long-Range Planning Committee as a standing committee to bring to the attention of Council items of importance to the Institute and the profession.

The Institute grew at a rapid rate. Chemical engineering enrollments increased significantly and employment opportunities exceeded supply. The important role of the chemical engineer, not only in the traditional areas, but in addressing energy and environmental problems, in electronics, in genetic engineering, and in food production, to name but a few, was being recognized. However, this situation brought about a crisis in the nation's departments of chemical engineering as burgeoning undergraduate enrollments, low graduate enrollments, depleted faculty ranks, and meager equipment budgets combined to severely strain the nation's engineering education apparatus. The Institute recognized the existence of this crisis early and in November, 1981, the joint efforts of several groups were combined into a Committee on Industry/Academic Coordination to define areas in which AIChE and its membership can take positive action to ease the crisis.

An exciting development was the formation of the Management and Marketing Divisions, hence providing an opportunity for chemical engineers in those areas to participate in AIChE activities. This recognizes the diversity of the chemical engineers and of the Institute's membership.

AIChE has always supported the concept of a single engineering organization to represent all engineers in the nation. Several presidents before me worked very hard with other Founder Society Presidents to organize an umbrella engineering organization to do those things for all professional engineers that individual societies cannot do alone. Finally, in late 1979 the American Association of Engineering Societies (AAES) was founded. It included large numbers of participating societies representing nearly a million U.S. engineers. A large number of AIChE members have participated in AAES activities. AAES (by the end of 1981) had not reached its full potential but AIChE is supportive of the concept and is working to make it successful.

During the period 1979-81 there was considerable increase in international activities in the form of joint meetings with individual chemical societies in other countries as well as participation with several foreign groups in international congresses. To coordinate these expanding activities a standing International Activities Committee was organized in 1981.

One cannot help but be optimistic about the future. The need for chemical engineers was never greater. The Institute grew rapidly and the 100,000 membership mark should be reached well before the Institute's 100th birthday. Then, as now, I am confident that AIChE will be at the forefront of activities relative to the betterment of society and the chemical engineering profession.

SEVENTY-FIVE YEARS OF PROGRESS

W. Kenneth Davis
—1981—

My recollections and impressions of the AIChE during the past 25 years include a variety of different kinds of things. Important among these are the development and reinforcement of many close personal friendships with truly first-class people during my service as a division chairman (the first AIChE Division, Nuclear Engineering), during three years of service on the Council at a time of considerable change, and lately (although not lastly) my three years on the Executive Committee as Vice President, President, and as the Past President or "fifth wheel." It is surprising in retrospect to realize that I really have spent a lot of the past quarter century deeply involved in the affairs of the Institute.

While the various meetings and affairs have been very pleasant and I deeply treasure many of the experiences at them, as does Margaret, I am most impressed by the unique character of chemical engineering and the way in which chemical engineers with the active intervention and support of the AIChE have met the broad and unusual challenges of our society during the past quarter century. In some ways it appears that chemical engineering was designed as a discipline to meet the opportunities presented by new requirements in the changing world of the 60's, 70's, and 80's.

New fields of importance to chemical engineering have included the increased need to prevent degradation of the environment, increased emphasis on public health and safety, the development of a whole new computer and related control technology, the emergence of biological and genetic engineering, the need to recover minerals and other valuable materials from ever depleting resources, often in remote locations, and a variety of others which scarcely existed 25 years ago. The evidence is overwhelming that this trend of new technological areas will continue into the future with perhaps even more emphasis on worldwide applications and developments. I am also confident that chemical engineering will continue to play a key role in these areas which are of vast importance to our civilization.

It has been my good fortune during this period to be involved both professionally and with the Institute in two key areas where chemical engineers have and continue to play important if not the key roles; nuclear energy and energy development. I believe a few comments on these areas will be useful in stressing the unusual character of chemical engineering in making important contributions to our society. While much of this could be treated as chronological history it seems to me that important points may be lost.

Nuclear technology was born during the enormous pressures of war and only became a real potential source of energy for civilian use during the 1950's. The key leaders during the wartime years were physicists and, as can be ascertained by reading any history of that period, chemical engineers. It seems clear that they had the basic education, the flexibility in thinking, and the imagination and ability to synthesize which allowed them to take on and successfully do the crash developments and projects needed. This included not only the chemical reprocessing where the contributions were obvious (although more difficult than might be supposed) but also the processing of uranium and plutonium, uranium enrichment, reactor design and operations, and many aspects of weapons design and production. When the opportunities for civilian use opened up the chemical engineers were and have continued at the forefront of this effort in almost every aspect of it.

The emergence of energy development as a major facet of our nation's priorities goes back well before the 1973 Arab oil embargo and the OPEC cartel price increases. Chemical engineering has been of direct application in such areas as oil shale, oil and gas from coal, enhanced oil recovery, improved refining, alternative petrochemical feedstocks and processes, and many others. Important contributions are being made in other less obvious areas such as photovoltaic development, biomass, methanol and ethanol production, etc. It seems likely that this emphasis will continue since we will find it necessary to develop and use new sources of energy.

History has demonstrated the enormous value of the broad based education in the fundamentals of chemistry, physics, mathematics, and materials combined with the concepts of unit processes and operations integrated into overall systems designed to operate economically, efficiently and safely. This has stimulated constructive thinking and problem solving in the real world.

One of the characteristics that has been invaluable to the chemical engineer has been the drive and ability to think ahead. The last quarter century has seen several planning studies and I am particularly pleased that the last three years has led to the establishment of a permanent, very active Long Range Planning Committee.

I believe we have many reasons to be proud of our profession and its accomplishments but have also succeeded in avoiding the deadly pitfalls of complacency and self-satisfaction. It seems to me that this is largely because of the type of people who have continued to lead the Institute.

SEVENTY-FIVE YEARS OF PROGRESS

Richard R. Hughes
—1982—

Most Institute presidents find their year active and diverse, and 1982 was no exception. The excellent work of the members and staff provides good momentum for most Institute activities, but there always seem to be new questions and new activities for the President.

As my special AIChE endeavor, I tried to strengthen the AIChE contribution to solving the engineering education crisis. A special committee of industrial and academic AIChE members was asked to review the many Institute projects related to education; its summary and action plan were presented to Council in November. AIChE also participated strongly in cooperative efforts with the other engineering societies, AAES, and other special groups. All of these efforts were aimed at solving the threatening inadequacy of educational facilities and engineering faculties in the face of growing demand for engineers, and increasing enrollment. Although it is hard to identify any specific improvements with our efforts, they did supply a stimulus and identify data for the development of industrial support, the improvement of educational programs, and the strengthening of government support for *engineering* education.

The long history of AIChE's interest in education was pointed out by our recognition as a Founding Member of ECPD at the ECPD/ABET 50th Anniversary celebration in October. By the time this meeting was held, it had become apparent that the engineering education crisis had taken a new turn. At most engineering colleges, there are still too few faculty and too many students, in spite of tighter admission requirements, and most educators feel this is a long-range problem, still awaiting solution. However, the turndown in the economy in mid-1983 added a new crisis. Suddenly it became difficult for the new graduates to find employment, adding to the problems caused by layoffs and early retirement of working engineers. The Council took the unprecedented step of establishing, at the Los Angeles November meeting, a first employment forum in the attempt to provide special support for engineers without jobs and encourage industry to strengthen their efforts to find places for these unemployed engineers.

In 1982, the unusual feature, for Mrs. Hughes and me, was its international flavor. The AIChE participation in the international scene promotes worldwide cooperation in applying chemical engineering, but it also has its, sometimes amusing, social and personal aspects.

Our international year began immediately following my inauguration at the November, 1981, New Orleans meeting. We joined the U.S. delegation, led by Chuck and Ursula Forman, to the meeting in Buenos Aires of the World Federation of Engineering Organizations. The conference wrestled with several serious (?!) policy questions, such as, "which Chinese delegation should be seated?" In this case, the compromise, found after numerous committee meetings, was to seat the People's Republic as the "national" delegation from China and the delegation from Taiwan as a "special member with all the rights and privileges of a national member," but without the title. The formal session used a large horseshoe arrangement; the U.S.A. delegation was seated between those from U.K. and U.S.S.R. The U.K. delegation was led by a baronet with a dry British wit, while the Soviet delegation included a young man wearing a leather jacket and dark glasses, who whispered continuously in his chief's ear. We felt very much "on the scene." Fortunately, the formal session was brief. More significant were the commission meetings, technical sessions, and contacts made at the magnificent (and crowded) social events. In August, we saw many of our Latin-American friends again at the UPADI meeting in San Juan, Puerto Rico. But the culmination of our international year was a trip around the world, made possible by consecutive meetings in China and London.

The China meeting originated with an invitation from the Chemical Industry and Engineering Society of China (CIESC). The American delegation of about 100 met for three days with 200 Chinese engineers in a large hotel on the outskirts of Peking. The presentations were in English, with a few exceptions where continuous translation was provided. In the opening plenary sesssion, I had an opportunity to reply to the welcoming address from Mme. Tao Tao, the President of the Chinese Institute. I emphasized the professional, democratic nature of AIChE, with strong membership participation in meetings and committees, and with policy set by a Council elected democratically in a *contested* election. To close the meeting, we all had the opportunity to attend a large banquet in the Great Hall of the People at Tien'anmen Square. Since our group only had 300 attendees, we were in one of the "smaller" banquet halls. (The large hall was being used to entertain Prime Minister Margaret Thatcher of the United Kingdom.) We were given first-rate treatment. The AIChE officers, the meeting chairmen, and session chairmen were invited with their spouses to a preliminary session. We all sat in comfortable chairs in a large meeting room. Elizabeth and I were led to seats immediately adjacent to those of the Minister of Chemical Industry and Mme. Tao Tao. The Minister then proceeded to speak into his microphone at some length. On translation, this proved to be some welcoming remarks, with the obvious implication that I should reply in the microphone immediately in front of me. I had been given no warning that this was to occur, but I managed to put together some intelligent phrases, and breathed a sigh of relief. But my ordeal wasn't over. While waiting for the translation, the Minister had apparently thought up some further comments. This sort of exchange went on for eight-to-ten exchanges, each comment lasting 1–2 minutes. I found it very difficult to find new things to say in this somewhat "stilted" conversation. However, the other members of the delegation and the U.S. Science Attache who were present reported that my *ad hoc* responses were effective and interesting. At the appropriate time the Minister rose and led us into the banquet hall to join the remainder of the U.S. delegation and guests. There, he immediately proposed the first toast, to which I responded. Several toasts were expected throughout the meal. The Minister and many of the other Chinese concentrated on maotai, a rather strong clear liquor. After one toast with that, I switched to the red wine. This was quite sweet, but at least not quite so potent. All of the Chinese toasts ended with the phrase "to the health of Dr. and Mrs. Hughes." After the joint meeting, most of the U.S. delegation spent from three to fourteen days touring Peking and other parts of China (at their own expense). We came away with a great admiration for the industry and ability of the Chinese; however, we also had a great appreciation of the problems which they face.

For Elizabeth and me, the trip continued around the world. After brief visits in India with former students, we participated in the Double Jubilee of the Institution of Chemical Engineers in London. Again, there was a strong American delegation, including most of the AIChE officers. The gift from AIChE of a commemorative pewter plate was extremely well received by the Institution. This visit might have been anti-climatic after the unique China meeting, but we were among old friends and subject to the warm British hospitality and service. A most fitting end to our international year.

Robert H. Marshall
—1983—

In our 75th Diamond Jubilee year the Institute is in excellent shape. We continue to grow at a good rate and expect to have more than 65,000 members by year end. We have a fine organization of outstanding volunteers. Our 11 divisions, more than 100 sections, about 135 student chapters and approximately 30 national committees are dynamic, active

ANECDOTES

groups. Our New York staff and Washington representatives continue to do an excellent job.

Many Institute groups have recommended to our Long Range Planning Committee that AIChE become more active in New Technology and innovation. A New Technology Committee of members who have a good track record in this area has been formed. They will give us guidance.

Although most Institute affairs are in good shape we do have several important problems which greatly concern us. One of the more critical is the one of unemployment of a small but very important part of our membership. The Institute in Los Angeles in 1982 and Houston in 1983 temporarily waived the prohibition on recruiting, and had successful Employment Clearinghouses at both meetings. Many sections have formed active Employment Committees and are aiding their unemployed members in seeking jobs.

Although we have had a small improvement this year as concerns the Crisis in Chemical Engineering Education, we continue to "eat our seed corn." The several problems of shortage of chemical engineering professors, low pay for professors and graduate students, few native born graduate students and, at some schools, inadequate and obsolete unit operations laboratory equipment are still with us. Many groups both within and outside of the Institute are working hard to solve these problems.

Anniversaries such as our 75th give a good excuse for reminiscing. Until the late 1950's my association with AIChE had been minimal. In the early 1940's I was a typical member of the LSU Student Chapter. In the mid-1940's as a member of a U.S. Army combat unit I was more concerned with keeping my head attached to my body; I did not place AIChE high on my priority list.

My early days in industry were spent travelling around much of the country for Ethyl Corp., so there was little time for AIChE, or any organization. In the late 1950's I met two of the real dynamos of the Institute, Johnny McKetta and Bill Cunningham of the University of Texas. It wasn't long until they had twisted my arm and got me involved in two AIChE activities, organizing several continuing education courses for the South Texas Section, and working on a committee for the section's Annual Technical Meeting. This led to a series of committee jobs for several Houston National Meetings.

In the mid-1970's the South Texas Section honored me with election as Vice Chairman and Chairman of the Institute's largest section.

In 1975 Bill Corcoran convinced me that I should become more involved in national activities. This led to election as a Director in 1978, participation on the Government Programs Steering Committee, chairing the *ad hoc* Continuing Professional Competence Committee and working with the formation committee for the new Safety and Health Division.

In 1981 you honored me with election as Vice President for 1982 and President for 1983.

We have, I believe, the most dynamic of the several engineering organizations. You, our volunteers, have done a great job for our members, and for the welfare of the nation. Let's keep up the good work and make the world a better place in which to live. In 2008 at our hundredth anniversary let's look back on the preceding 25 years and be able to say "well done".

Stephen L. Tyler
Secretary (1937–54)

As I See It

The future of chemical engineering is a subject I have chosen to write about, on the occasion of my retiring after eighteen years as secretary of the Institute. My remarks will, on the whole, be optimistic. In this respect I have elected to differ from many of my non-chemical engineering contemporaries, by not taking a gloomy view of the future. There is a very good reason for this; I am of the belief that chemical engineering is going to provide the answers to a number of the problems and desires that a great many people worry about. This will not be accomplished, however, without overcoming problems within the profession, and I want also to discuss these.

The general outlook for chemical engineering is excellent. I base this conclusion on employment opportunities, professional status development, and the prospects for the future of the chemical process industries. In this latter connection, I cannot see any ceiling over the construction of new processing facilities. Prospects for successful commercialization of the hundreds or even thousands of untried organic chemicals, pharmaceuticals and plastics have never been better. The public is educated to expect wonders from commercial syntheses, and buys eagerly those products which add so much to our modern way of life. On the other hand improvements are constantly being made in other existing products and processes, maintaining a high rate of obsolesence of facilities for processing.

Automation in the process industries is already making its demands on the chemical engineer, who is rightfully expected to judge the over-all effectiveness of control function application, just as he is best equipped to judge any other factors affecting the technology and economics of the process plant. To keep up with advances in control devices, the chemical engineer will find it necessary to develop a close acquaintance with what can be expected from the multitude of available choices without departing from his basic role of materials-processor.

ANECDOTES

The nuclear age is depending on the chemical engineer to apply his techniques to the solution of one of the major problems in the way of economic nuclear power. This has to do with commercial scale separation of neutron-absorbing byproducts formed in fuel as it is used. To remove these, such familiar unit operations as ion exchange and solvent extraction are being used successfully in small scale, but with hideous difficulties arising from radiation hazard and corrosion ready to confront the large scale operator.

Along with having to learn much more about his technology, both applied and theoretical, the chemical engineer is being looked to for greater participation in affairs relating to government, community, and human relations. He is very much wanted in management, as well as in conducting negotiations resulting in the sale of materials, equipment and services used in processing. I take great pride in seeing the progress being made by the Institute in better serving its members' growing needs through the inclusion of a broader range of subject material in its publishing program and through the activities of its Professional Development Committee. There is every reason to believe we are at the threshold of a new era in the professional society's service to the engineer.

Of course, there aren't enough chemical engineers. To some, this may appear as conducive to personal prosperity. I believe, however, we are so far from a state of saturation, that we may discard our fears and concentrate instead on finding better ways to make use of the chemical engineer's talents. We might as members of society take serious note of the graduation rate of engineers and scientists in the Soviet Union, and ask if we have not yet reached the point where we can expect Selective Service to endeavor to better appreciate the worth of our engineers working as civilians. The conservation of our chemical engineering manpower resources might be aided by placing greater emphasis on the training of technical aides.

In closing, let me say that happiness with his position is within the province of every chemical engineer, and that it is a prime requisite to success.

Steve Tyler

SEVENTY-FIVE YEARS OF PROGRESS

Franklin J. Van Antwerpen
Secretary (1955–78)

A Final Accounting of Stewardship

Walk worthy of the vocation wherewith ye are called.

EPH.4:1

Retiring is funny business: you know what you are giving up, but you don't know what you are getting into. Like now. Since I am about to retire, laying down my AIChE burdens, I thought it would be good to wind up with a final column on my stewardship. Every year end, since 1955, when I became Secretary and Executive Secretary (changed now to Executive Director), I wrote a column called An Accounting of Stewardship; so why not do the same for the whole period? A month ago, even three days ago, it seemed like the thing to do. But now that I am faced with writing that final accounting and the words aren't coming easily and the clock insists that I produce copy, it isn't such a good idea after all. What was the essence of my AIChE career? I can't really say, and I'll not try. But rather than go unspoken into senior citizenry, I can offer my thanks.

First, to the members, who make AIChE possible. There is something awfully right about the type of person in chemical engineering when our professional society has a greater percentage of the chemical engineers than any other engineering society has of its possible members. Keep that hallmark through the long future.

And thanks for selecting Directors and Officers who are leaders and who are willing to sacrifice to the ideal of being a professional. Serving AIChE costs much personal time—evenings and weekends. It costs money: our Directors pay their own way or their company underwrites their expenses, but either way it's a contribution to chemical engineering. Being a Director or Officer requires a willingness to shoulder still one more load, to write another report, to phone another chairman, to rewrite that "another report," to take one more post. AIChE needs leaders who are not deterred by criticism. Keep electing chemical engineers just like you: capable, fair, and ready to help.

And my special praise and thanks for AIChE Presidents. One gets to more than admire each one during that year in which their leadership qualities are put to the test. Their judgments must be quick and accurate; their tempers quiescent; and their powers such as to graciously weld members, committeemen, and Directors into a happy, dedicated team. All my Presidents were special gems, and all of us are lucky they cared enough to serve.

And then the AIChE staff. I almost said my staff, but they are yours. AIChE's successes—and there have been many in those 32 years—belong largely to them. They have rejoiced quietly along with me at each step upward, relishing what was accomplished but looking to the next step. And they faced our failures quietly also, analytically always: next time the pitons were driven deeper, the route was more carefully plotted, the crevices were bridged more sturdily. But for most of the way there has been that happy last dash to the top—all together—each doing a job, each an individual, each an expert.

To all, then, members, Directors, Officers, Presidents, staff—my great thanks, my great love. It was fun. I bless you all.

My thanks, too, to those who are especially important to me—my family. My parents, now gone, I remember joyously because of their courage. I remember their sacrifices. My two children, Franklin and Virginia, in being children taught me how to be a father and now, in being adults, have given me great pride.

And Dorothy—which means gift of God, and she is—has been especially important! This private dialogue from 1946:

"The AIChE has asked me to begin a publications program. I will have to take a cut in salary, they have no pension plan, no hospitalization—nothing, not an office, not a staff—just 'Van, we want you to start a publications program for us.' "

After a few seconds of silence, "Would you want to do it?"

"Well, I am a chemical engineer, and AIChE sure needs a publications program right now. But it's an awful lot to ask—a big risk."

"Dear, go ahead and do it. Money and security we can get any time. But how many men ever get an opportunity to start a new publication for their profession?"

As I said before: Thanks.

F.J.V.A.

SEVENTY-FIVE YEARS OF PROGRESS

CHAPTER 6

A Look to the Future

J. Charles Forman
Executive Director &
Secretary, AIChE

A Jubilee Celebration is an opportune time to look forward as well as back. Terry Reynolds' history, I'm sure all will agree, has captured the flavor of the growth of AIChE over the past 75 years. Starting from a small and rather unsteady base, we have grown to the position of strength and respect that we now command. As you have read in this history, AIChE's growth has been steady, but not without some excitement and no small amount of discussion and controversy.

Sitting in 1983, attempting to look to the future (and realizing that the next time for a formal evaluation of AIChE's activities in the form of a history will probably be during our centennial in 2008) gives one thought; the early years of the decade of the 1980s are not the best times for the United States, the world economy, or for chemical engineering. The smooth growth era of the chemical and petroleum industries ended rather abruptly with the onset of the recession in the early 80s, and the outlook is for little if any growth in the traditional "smokestack" industries (such as steel, automobiles, and heavy chemicals) that built the American industrial economy.

As a result of the recession and corresponding increase in unemployment of AIChE members, Council in 1982 authorized several expanded activities in the area of services to unemployed members. First of all, the Professional Development Committee (now charged with professional employment activities) surveyed practices of companies that were forced to terminate members' employment; this included visits to interview selected companies regarding their use of the Employment Guidelines. Second, Council temporarily suspended AIChE's long standing rule against recruiting at our meetings by voting to set up employment clearing houses, starting at the 1982 Annual Meeting and continuing for the duration of the employment downturn. In addition, the crisis in engineering education and chemical engineering education in particular, is continuing with overcrowded engineering schools, and an aging faculty and physical plant. Coupled

A LOOK TO THE FUTURE

with the well publicized problems in primary and secondary education, especially in the sciences, this educational problem requires immediate attention at all levels of our government and society if the United States is to find the new directions needed to revitalize and reindustrialize our economy. Fortunately, AIChE has the organization, the funds, and most important, the tradition and will to act and lead in doing what needs to be done to contribute to the solution of these problems. We are the oldest, largest, and most active chemical engineering society in the world, well heard and respected, and thus uniquely qualified to move forward. We can and will do this because we have an excellent volunteer organization, the mechanisms (by that I mean the means we have set up to accomplish our goals) and, finally, the staff to do the job.

AIChE's volunteer organization is, as has been pointed out by Terry Reynolds in this history volume, one that is still somewhat unique amongst the large American engineering societies. Perhaps because we are the youngest of the major societies, we still operate with a strong Council, all of whose members are elected at-large and in contested elections. The Council, working with the Executive Director, sets the basic policies and directions for AIChE, and it has been a source of pleasure and pride for me to witness the workings of the Council over the past five years, to see how well your Council deliberates and acts for the good of AIChE and the profession. We have a strong and growing number of divisions to meet the future technical demands of our members.

With over 100 local sections, we have a section in most areas of the country that have enough chemical engineers to maintain a viable activity. Like so many other facets of modern day American life, local section activity, has become a less important part in the life of most American chemical engineers than it was at the time of our Golden Jubilee 25 years ago. The demands upon a chemical engineer's time are such that today a lower percentage of section members attend meetings and take part in other activities than in the 1950s and 1960s; however, local sections are still the backbone of any technical society and we are proud that the strong and active sections in AIChE are able to contribute in the positive way they do to our forward progress.

Our national committees are of course constantly changing in scope and activity, as befits a dynamic organization. For example, within the last two years our national Program Committee has completely reorganized itself to better categorize subjects for AIChE programming; has recommended and had Council approve a new policy on meeting size, to meet the expanding demands for chemical engineering programming; and has established and stressed with great success the "block programming" concept, in which multi-session program blocks are set up at each AIChE meeting to enable members to attend a full series of technical sessions in their particular areas of specialization. As the world's largest chemical engineering organization, we take part in many joint meetings with other societies, both here and abroad. Finally, our international contacts, both with our sister societies in other countries and through the established area federations of chemical engineering, keep us well informed on chemical engineering activities around the world. If America is to maintain its redirection and grow into new technology areas, we must know what our colleagues are doing in other countries.

The mechanisms to accomplish our goals are functioning and in place. Many of them are mentioned above in our strong volunteer organization. Many activities that are just starting around the time of this Diamond Jubilee celebration should serve us well over both the near and longer term future. Chief among these is our present activity in long range planning. The new Long Range Planning Committee established by Council in 1980 worked throughout 1981 and 1982 in developing its first report. The first phase of the long range planning study consisted of getting "feedback" from the various organizations in the Institute and their leaders. This feedback and the report were organized into a discussion of eight planning issues posed for the Institute, as follows:

1. What is chemical engineering, and what is a chemical engineer?
2. What will be the make-up of the chemical engineering profession and AIChE ten years or more down the road?
3. What will be the impact of new technologies on chemical engineering and AIChE?
4. The Structure of AIChE.
5. Communications.
6. Engineering Education.
7. Government Activities.
8. Societal Problems.

The first report of the committee was submitted to Council in 1983, and action was already underway by the time of the Diamond Jubilee Annual Meeting. As a matter of fact, the official definition of chemical engineering was modified by the committee so as to broaden its scope. It was approved by Council, submitted to the Members and Fellows of the Institute and was approved early in 1983. The new definition is as follows: *Chemical engineering is the profession in which a knowledge of mathematics, chemistry and other natural sciences gained by study, experience and practice is applied with judgment to develop economic ways of using materials and energy for the benefit of mankind.* The next stage of the long range planning study will probably be more of a strategic nature, with investigations of the economic future and status of the basic industries employing chemical engineers with recommendations for new directions both for the country and the Institute.

Among the new "mechanisms" has been the formation of a New Technology Committee (one of the Long Range Planning Committee recommendations) to investigate and recommend new areas of interest for chemical engineers; this, of course will usually show up first in our meeting programs, publications, etc. As an example of this, three new quarterly publications in the *Progress* group are undoubtedly only the start of new publication ventures by AIChE into new technology areas. By the time of the Diamond Jubilee, new publications in such areas as computers, bioengineering, and others were already under discussion.

Finally, the job of the AIChE headquarters staff is to implement the directives of Council and to work for the membership and the profession. When I became Executive Director in 1978, one of my first steps was to reorganize the headquarters staff along functional lines; we now have the basic five staff directorates. We have taken over the management of our expositions in-house, for better management, closer control, and appreciable cost savings. We have added new equipment and adopted new tools to work more efficiently, especially the better use of computers and data processing and word processing equipment. We now have our member and other records available on key employees' desks on video display terminals, do all our meeting registration on a portable mini-computer, and are looking for new, better, and faster tools for the future.

In the year 2008, on the occasion of the AIChE Centennial celebration, the centennial history will detail the events of the quarter-century between now and then and may well have another "look to the future". I'm looking forward to working with AIChE and its members over the next decade or so, and reading the centennial history with interest and pride.

Appendix

OFFICERS, 1908–1983

PRESIDENTS

R. H. Marshall	1983	Thomas H. Chilton	1951
R. R. Hughes	1982	Warren L. McCabe	1950
W. Kenneth Davis	1981	Francis J. Curtis	1949
J. G. Knudsen	1980	Albert B. Newman	1948
J. Y. Oldshue	1979	Charles M. A. Stine	1947
W. H. Corcoran	1978	James G. Vail	1946
A. S. West	1977	Lawrence W. Bass	1945
K. D. Timmerhaus	1976	George Granger Brown	1944
K. E. Coulter	1975	J. L. Bennett	1943
Irving Leibson	1974	Sidney D. Kirkpatrick	1942
Theodore Weaver	1973	Francis C. Frary	1941
T. W. Tomkowit	1972	Webster N. Jones	1939–40
J. J. Martin	1971	Fred C. Zeisberg	1938
A. L. Conn	1970	Martin H. Ittner	1936–37
H. D. Guthrie	1969	Albert E. Marshall	1934–35
M. S. Peters	1968	John V. N. Dorr	1932–33
T. A. Burtis	1967	John C. Olsen	1931
S. W. Churchill	1966	Alfred Holmes White	1929–30
W. B. Franklin	1965	E. R. Weidlein	1927–28
D. A. Dahlstrom	1964	Hugh K. Moore	1926
W. R. Marshall	1963	Charles L. Reese	1924–25
J. J. McKetta	1962	Henry Howard	1922–23
J. J. Healy	1961	David Wesson	1920–21
Jerry McAfee	1960	Arthur D. Little	1919
Donald L. Katz	1959	G. W. Thompson	1917–18
G. E. Holbrook	1958	G. D. Rosengarten	1915–16
J. Henry Rushton	1957	M. C. Whitaker	1914
Walter G. Whitman	1956	T. B. Wagner	1913
Barnett F. Dodge	1955	L. H. Baekeland	1912
Chalmer G. Kirkbride	1954	F. W. Frerichs	1911
William T. Nichols	1953	Charles F. McKenna	1910
William I. Burt	1952	Samuel P. Sadtler	1908–09

VICE-PRESIDENTS

J. H. Sanders	1983	Francis J. Curtis	1948
R. H. Marshall	1982	Albert B. Newman	1947
R. R. Hughes	1981	Charles M. A. Stine	1946
W. Kenneth Davis	1980	Albert B. Newman	1945
J. G. Knudsen	1979	Lawrence W. Bass	1944
J. Y. Oldshue	1978	George Granger Brown	1943
W. H. Corcoran	1977	James L. Bennett	1942
A. S. West	1976	Sidney D. Kirkpatrick	1941
K. D. Timmerhaus	1975	James G. Vail	1939-40
K. E. Coulter	1974	Webster N. Jones	1938
Irving Leibson	1973	Fred C. Zeisberg	1936-37
Theodore Weaver	1972	Harry A. Curtis	1934-35
T. W. Tomkowit	1971	Albert E. Marshall	1932-33
J. J. Martin	1970	John V. N. Dorr	1931
A. L. Conn	1969	John C. Olsen	1929-30
H. D. Guthrie	1968	A. H. White	1928
M. S. Peters	1967	R. T. Haslam	1927
T. A. Burtis	1966	John V. N. Dorr	1926-27
S. W. Churchill	1965	H. S. Miner	1922-26
W. B. Franklin	1964	H. K. Moore	1920-25
D. A. Dahlstrom	1963	A. H. White	1924
W. R. Marshall	1962	A. W. Smith	1919-23
J. J. McKetta	1961	Henry Howard	1917-21
J. J. Healy	1960	John M. Stillman	1915-19
Jerry McAfee	1959	A. C. Langmuir	1914-18
D. L. Katz	1958	G. W. Thompson	1913-16
George E. Holbrook	1957	R. K. Meade	1913-14
J. Henry Rushton	1956	M. C. Whitaker	1912-13
Arthur K. Doolittle	1955	H. B. Wagner	1912
Barnett F. Dodge	1954	Eugene Haanel	1910-12
Chalmer G. Kirkbride	1953	George P. Adamson	1911
William T. Nichols	1952	L. H. Baekeland	1911
William I. Burt	1951	F. W. Frerichs	1910
Thomas H. Chilton	1950	Edward G. Acheson	1908-10
Warren L. McCabe	1949	H. A. Hunicke	1908-09
Charles F. McKenna		1908-09	

DIRECTORS

D. B. Nelson	1983-85	R. F. Anderson	1981-83
J. D. Seader	1983-85	H. T. Brown	1981-83
L. O. Stine	1983-85	L. B. Evans	1981-83
A. A. Winkler	1983-85	H. S. Kemp	1981-83
L. F. Albright	1982-84	James A. Buckham	1980-82
F. J. Endelman	1982-84	Lee C. Eagleton	1980-82
E. R. H. McDowell	1982-84	John P. Sachs	1980-82
S. I. Proctor, Jr.	1982-84	John H. Sanders	1980-82

APPENDIX

B. C. Doumas	1979-81	J. J. Martin	1967-69
T. H. Goodgame	1979-81	J. L. Olsen	1967-69
H. M. Rodekohr	1979-81	E. B. Christiansen	1966-68
C. R. Vander Linden	1979-81	A. L. Conn	1966-68
F. A. M. Buck	1978-80	F. C. Croxton	1966-68
H. W. Flood	1978-80	W. W. Ellis	1966-68
R. H. Marshall	1978-80	J. R. Fair	1965-67
E. A. Weinman	1978-80	H. D. Guthrie	1965-67
L. L. Fellinger	1977-79	R. L. Pigford	1965-67
S. E. Isakoff	1977-79	Alfred Smith, III	1965-67
P. H. McNamara	1977-79	W. W. Kraft	1964-66
M. P. Skillern	1977-79	E. M. Schoenborn	1964-66
J. J. Fox	1976-78	Mott Souders	1964-66
A. E. Humphrey	1976-78	A. S. West	1964-66
R. L. Jacks	1976-78	T. A. Burtis	1963-65
B. I. MacDonald	1976-78	W. M. Carlson	1963-65
J. B. Butt	1975-77	L. J. Coulthurst	1963-65
A. V. Caselli	1975-77	B. B. Kuist	1963-65
C. C. Neas	1975-77	Stuart W. Churchill	1962-64
J. W. Prados	1975-77	John W. Clegg	1962-64
J. A. Babcock	1974-76	Wayne C. Edmister	1962-64
D. M. Himmelblau	1974-76	H. B. Warner	1962-64
H. L. Hoffman	1974-76	Robert J. McNally	1961-63
J. G. Knudsen	1974-76	Max S. Peters	1961-63
C. S. Cronan	1973-75	Carl F. Prutton	1961-63
G. A. Lessells	1973-75	Robert R. White	1961-63
W. R. Pavelchek	1973-75	Donald A. Dahlstrom	1960-62
W. A. Schnyder	1973-75	Karl H. Hachmuth	1960-62
K. B. Bischoff	1972-74	W. Kenneth Menke	1960-62
W. Kenneth Davis	1972-74	Charles R. Wilke	1960-62
W. R. Earley	1972-74	W. R. Collings	1959-61
K. D. Timmerhaus	1972-74	W. B. Franklin	1959-61
C. H. Chilton	1971-72	L. C. Kemp, Jr.	1959-61
W. H. Corcoran	1971-73	Roland Voorhees	1959-61
K. E. Coulter	1971-73	Raymond P. Genereaux	1958-60
R. B. Filbert, Jr.	1971-73	E. R. Gilliland	1958-60
E. O. Ohsol	1973	John J. McKetta	1958-60
E. L. Ekholm	1970-72	Henry F. Nolting	1958-60
J. Y. Oldshue	1970-72	Manson Benedict	1957-59
James Wei	1970-72	John J. Healy	1957-59
R. E. White	1970-72	Joseph H. Koffolt	1957-59
R. C. Reid	1969-71	Richard H. Wilhelm	1957-59
L. W. Leggett, Jr.	1969-71	W. R. Marshall, Jr.	1956-58
R. R. Hughes	1969-71	Jerry McAfee	1956-58
W. B. Howard	1969-71	D. F. Othmer	1956-58
J. B. Martin	1968-70	C. A. Stokes	1956-58
T. W. Tomkowit	1968-70	E. B. Chiswell, Jr.	1955-57
Theodore Weaver	1968-70	W. A. Cunningham	1955-57
J. W. Westwater	1968-70	D. L. Katz	1955-57
G. H. Cummings	1967-69	W. E. Lobo	1955-57
Irving Leibson	1967-69		*(continued next page)*

APPENDIX

Directors continued

Name	Years	Name	Years
Ray P. Dinsmore	1954-56	Thomas H. Chilton	1942-44
W. L. Faith	1954-56	Warren L. McCabe	1942-44
George E. Holbrook	1954-56	Mark E. Putnam	1942-44
Loren P. Scoville	1954-56	Lawrence W. Bass	1943
William T. Dixon	1953-55	Francis J. Curtis	1943
Eger V. Murphree	1953-55	Chester L. Knowles	1943
George G. Oberfell	1953-55	Norman W. Krase	1941-43
Lee Van Horn	1953-55	Albert E. Marshall	1941-43
Arthur K. Doolittle	1952-54	Robert L. Murray	1941-43
Joseph C. Elgin	1952-54	James G. Vail	1941-43
Philip H. Groggins	1952-54	W. G. Whitman	1942
J. Henry Rushton	1952-54	George G. Brown	1940-42
William T. Dixon	1952	Charles R. Downs	1940-42
Ralph H. Price	1952	Albert B. Newman	1940-42
Robert C. Gunness	1951-53	John H. Perry	1940-42
R. Paul Kite	1951-53	James Leroy Bennett	1939-41
Charles R. Nelson	1951-53	Barnett F. Dodge	1939-41
Earl P. Stevenson	1951-53	Victor L. King	1939-41
G. E. Holbrook	1950-52	Charles M. A. Stine	1939-41
C. G. Kirkbride	1950-52	Donald B. Keyes	1939-40
R. L. Murray	1950-51	Lawrence W. Bass	1938-40
W. G. Whitman	1950-51	H. A. Curtis	1938-40
Donald B. Keyes	1949-51	Francis C. Frary	1938-40
Paul D. V. Manning	1949-51	Thomas H. Chilton	1937-39
Irvin L. Murray	1949-51	Gaston F. DuBois	1937-39
Henry D. Wilde	1949-51	Sidney D. Kirkpatrick	1937-39
William I. Burt	1948-50	W. G. Whitman	1937-39
J. J. Healy, Jr.	1948-50	Victor L. King	1938
M. C. Molstad	1948-50	Ellery L. Wilson	1938
William T. Nichols	1948-50	Edward Bartow	1936-38
Thomas H. Chilton	1947-49	Gregory Egloff	1936-38
Joseph C. Elgin	1947-49	Albert B. Newman	1936-38
Sidney D. Kirkpatrick	1947-49	James G. Vail	1936-38
Thomas K. Sherwood	1947-49	Lawrence W. Bass	1937
C. G. Kirkbride	1946-48	James Leroy Bennett	1935-37
Warren L. McCabe	1946-48	Robert B. Harper	1935-37
John H. Perry	1946-48	Webster N. Jones	1935-37
J. K. Roberts	1946-48	Victor L. King	1936
A. P. Colburn	1945-47	Gustavus J. Esselen	1934-36
Guy N. Harcourt	1945-47	Albert G. Peterkin	1934-36
Francis J. Curtis	1945-47	Ellery L. Wilson	1934-36
C. C. Furnas	1945-47	J. R. Withrow	1934-36
William I. Burt	1944-46	F. C. Zeisberg	1935
James C. Lawrence	1944-46	Charles R. Downs	1933-35
William T. Nichols	1944-46	Harrison E. Howe	1933-35
David E. Pierce	1944-46	Sidney D. Kirkpatrick	1933-35
W. W. Duecker	1945	Frederic W. Willard	1933-35
Chester L. Knowles	1945	Willard H. Dow	1932-34
Barnett F. Dodge	1943-45	R. T. Haslam	1932-34
Charles M. A. Stine	1943-45	H. C. Parmelee	1932-34

APPENDIX

James G. Vail	1932-34	Frank Hemingway	1920-21
Ellery L. Wilson	1932-33	H. S. Miner	1920-21
James Leroy Bennett	1931-33	John V. N. Dorr	1919-21
H. A. Curtis	1931-33	Herbert Dow	1919-21
Webster N. Jones	1931-33	F. M. deBeers	1920
Crosby Field	1931-32	A. C. Langmuir	1920
A. A. Backus	1930-32	T. B. Wagner	1918-20
Gustavus J. Esselen	1930-32	Hugh K. Moore	1919
F. C. Zeisberg	1930-32	A. W. Smith	1918-19
Robert E. Wilson	1931	David Wesson	1918-19
W. L. Badger	1929-31	L. H. Baekeland	1917-19
Charles R. Downs	1929-31	M. H. Ittner	1917-19
L. V. Redman	1929-31	J. R. Withrow	1917-19
Frederic W. Willard	1929-31	Arthur D. Little	1916-18
W. K. Lewis	1930	Charles F. McKenna	1916-18
John V. N. Dorr	1928-30	Richard K. Meade	1916-18
W. C. Geer	1928-30	G. P. Adamson	1915-17
R. T. Haslam	1928-30	J. B. F. Herreshoff	1915-17
J. R. Withrow	1928-30	Samuel P. Sadtler	1915-17
Albert E. Marshall	1928-29	J. C. Hebden	1914-16
N. K. Chaney	1927-29	Henry Howard	1914-16
H. A. Curtis	1927-29	I. P. Lihme	1914-16
D. D. Jackson	1927-29	Thomas Griswold, Jr.	1915
F. C. Zeisberg	1928	Jokichi Takamine	1913-15
A. H. Hooker	1926-28	J. R. Withrow	1913-15
Harrison E. Howe	1926-28	G. D. Rosengarten	1913-14
S. W. Parr	1926-28	A. Bement	1912-14
Albert E. Marshall	1927	A. C. Langmuir	1912-14
J. C. Olsen	1927	H. S. Miner	1912-14
F. C. Zeisberg	1925-27	E. G. Acheson	1911-13
H. C. Parmelee	1925-26	W. M. Booth	1911-13
E. R. Weidlein	1925-26	Edward Hart	1911-13
W. L. Badger	1924-26	John T. Baker	1911-12
F. A. Lidbury	1924-26	W. M. Grosvenor	1910-12
Albert E. Marshall	1924-26	R. K. Meade	1910-12
Edward Bartow	1923-25	H. F. Brown	1909-11
John V. N. Dorr	1923-25	Ludwig Reuter	1909-11
Ralph H. McKee	1923-25	Thorn Smith	1909-11
E. R. Weidlein	1924	Samuel P. Sadtler	1910
William H. Bassett	1922-24	G. P. Adamson	1908-10
A. C. Langmuir	1922-24	Edward Gudeman	1908-10
Alfred H. White	1922-23	David Wesson	1908-10
F. E. Dodge	1921-23	J. M. Camp	1908-09
A. H. Hooker	1921-23	Charles A. Catlin	1908-09
W. D. Richardson	1921-23	Eugene Haanel	1908-09
Albert E. Marshall	1922	H. F. Brown	1908
C. T. Bragg	1920-22	Ludwig Reuter	1908
C. L. Reese	1920-22	Thorn Smith	1908

SECRETARIES

J. Charles Forman	1978–	Frederic J. Le Maistre	1930–March 1937
F. J. Van Antwerpen	1955–78	H. C. Parmelee	1927–29
Stephen L. Tyler	April 1937–54	John C. Olsen	1908–26

TREASURERS

J. Y. Oldshue	1983–	J. Henry Rushton	1958–62
E. A. Weinman	1981–82	G. G. Brown	July 1953–August 1957
W. R. Marshall	1976–80	C. R. De Long	1936–June 1953
A. S. West	1973–75	Martin H. Ittner	1925–35
T. A. Burtis	1970–72	F. W. Frerichs	1912–25
G. E. Holbrook	1963–69	H. S. Renaud	1911
	W. M. Booth	1908–10	

CHARTER MEMBERS OF THE INSTITUTE

Edward G. Acheson	F. W. Frerichs	Richard K. Meade
George P. Adamson	W. M. Grosvenor	A. L. Miller
Jerome Alexander	Edward Gudeman	Louis A. Olney
Lucius E. Allen	Eugene Haanel	John C. Olsen
G. E. Barton	George M. Heath	Charles Lee Reese
William H. Bassett	Herbert Hollick	Henry Stanley Renaud
A. Bement	David Wilbur Horn	Ludwig Reuter
William Miller Booth	Henry August Hunicke	Andrew Robertson
H. F. Brown	Walter Renton Ingalls	Samuel P. Sadtler
J. M. Camp	H. M. Kaufmann	Thorn Smith
Charles A. Catlin	Arthur C. Langmuir	A. P. Trautwein
Frederic Dannerth	Wm. P. Mason	David Wesson
Allan W. Dow	C. F. McKenna	J. Edward Whitfield
	Ferdinand G. Wiechmann	

NATIONAL COMMITTEES

Admissions (1908)
Awards (1946)
Career Guidance (1952)
Chemical Engineering Education (1908)
Chemical Engineering Projects (1938)
Constitution and Bylaws (1932)
Continuing Education (1963)
Education and Accreditation (1922)
Equipment Testing Procedures (1944)
Ethics (1911)
Executive (1936)
Finance (1908)
Government Programs Steering (1965)
International Activities (1980)

Long Range Planning (1980)
Membership (1950)
Minority Affairs Coordinator
Nominating (1954)
Personnel Supply and Demand (1979)
Professional Development (1935)
Professional Legislation (1930)
Program (1944)
Publication (1908)
Public Relations (1930)
Research (1950)
Standards (1954)
Student Chapters (1929)
Technicians' Affairs (1971)

APPENDIX

INSTITUTE DIVISIONS

Computing and Systems Technology (1977)
Environmental (1970)
Food, Pharmaceutical, and Bioengineering (1966)
Forest Products (1968)
Fuels and Petrochemicals (1966)
Heat Transfer and Energy Conversion (1956)
Management (1980)
Marketing (1980)
Materials Engineering and Sciences (1968)
Nuclear Engineering (1954)
Safety and Health (1979)

DESIGN INSTITUTES

DIMP—Multiphase Processing
DIERS—Emergency Relief Systems
DIPPR—Physical Property Data

AWARDS

William H. Walker Award

1936—Allan P. Colburn
1937—Thomas Bradford Drew
1938—Warren L. McCabo
1939—George Granger Brown
1940—Walter L. Badger
1941—Thomas K. Sherwood
1942—W. Clifford Williams
1943—H. F. Johnstone
1944—O. A. Hougen
1945—Hoyt C. Hottel
1946—John H. Walthall and Philip Miller
1947—Manson Benedict and L. C. Rubin
1948—Kenneth M. Watson
1949—W. H. McAdams
1950—Barnett F. Dodge
1951—Richard H. Wilhelm
1952—John Henry Rushton
1953—W. R. Marshall
1954—E. R. Gilliland
1955—Edgar L. Piret
1956—Edward W. Comings
1957—Joseph Clifton Elgin
1958—Robert L. Pigford

1959—Bruce H. Sage
1960—Joe Mauk Smith
1961—Neal R. Amundson
1962—Robert Byron Bird
1963—Robert E. Treybal
1964—Thomas J. Hanratty
1965—Charles R. Wilke
1966—James W. Westwater
1967—John M. Prausnitz
1968—Donald L. Katz
1969—Stuart W. Churchill
1970—Arthur B. Metzner
1971—Theodore Vermeulen
1972—Harry G. Drickamer
1973—J. R. Fair
1974—Leon Lapidus
1975—Edwin N. Lightfoot, Jr.
1976—C. Judson King
1977—L. W. Seriven
1978—Cedomir Sliepeevich
1979—Sheldon K. Friedlander
1980—James Wei
1981—J. L. Duda and J. S. Vrentas
1982—William R. Schowalter

(Awards continued next page)

Awards continued

Allan P. Colburn Award

1945—Clyde McKinley and Robert R. White
1946—Charles E. Lapple
1947—Harry G. Drickamer
1948—J. Edward Vivian and Roy P. Whitney
1949—E. G. Scheibel
1950—F. M. Tiller
1951—Charles R. Wilke
1952—Thomas Baron and Lloyd G. Alexander
1953—LeRoy Alton Bromley
1954—Harold P. Grace
1955—William E. Ranz
1956—Theodore Weaver
1957—Thomas Joseph Hanratty
1958—Arthur B. Metzner, Robert D. Vaughn and George L. Houghton
1959—S. K. Friedlander
1960—L. W. Scriven and Charles V. Sternling
1961—George R. Moore
1962—John N. Prausnitz
1963—Andreas Acrivos
1964—Herbert L. Toor
1965—Byron C. Sakiadis
1966—John A. Quinn
1967—Earl B. Adams and John C. Whitehead
1968—John B. Butt
1969—Edward F. Leonard
1970—Ronald B. Root and Roger A. Schmitz
1971—Dale F. Rudd
1972—Dan Luss
1973—C. A. Eckert
1974—William J. Ward, III
1975—Edward L. Cussler
1976—John H. Seinfeld
1977—Clark Kenneth Colton
1978—L. Gary Leal
1979—James E. Bailey
1980—Thomas F. Edgar
1981—W. Henry Weinberg
1982—George Stephanopoulos

Professional Progress Award in Chemical Engineering

1948—Allan P. Colburn
1949—Mott Souders, Jr.
1950—Edwin R. Gilliland
1951—Chalmer G. Kirkbride
1952—Richard H. Wilhelm
1953—George E. Holbrook
1954—John Ridgway Bowman
1955—Robert L. Pigford
1956—Robert Roy White
1957—Vladimir Haensel
1958—W. Kenneth Davis
1959—W. R. Marshall, Jr.
1960—William Gardener Pfann
1961—Thomas Baron
1962—Jack A. Gerster
1963—George E. P. Box
1964—Stuart W. Churchill
1965—Robert Byron Bird
1966—Leon Lapidus
1967—Thomas J. Hanratty
1968—Andreas Acrivos
1969—Cornelius J. Pings
1970—James Wei
1971—John S. Bonner
1972—Arthur E. Humphrey
1973—P. L. T. Brian
1974—Julian Szekely
1975—John H. Sinfelt
1976—Kenneth B. Bischoff
1977—Morton M. Denn
1978—John B. Butt
1979—Dan Luss
1980—L. Louis Hegedus
1981—K. A. Smith
1982—W. Harmon Ray

APPENDIX

Institute Lecture

1949—W. H. McAdams
1950—Olaf A. Hougen
1951—T. B. Drew
1952—W. R. Marshall, Jr.
1953—G. G. Brown
1954—Manson Benedict
1955—John G. Roberts
1956—W. K. Lewis
1957—C. R. Wilke
1958—Neal R. Amundson
1959—Thomas H. Chilton
1960—Joel O. Hougen
1961—Michel Boudart
1962—Hugh M. Hulburt
1963—Mott Souders
1964—J. W. Westwater
1965—Charles V. Sternling

1966—R. L. Pigford
1967—Robert C. Reid
1968—James Wei
1969—W. W. Akers
1970—J. Henry Rushton
1971—Judson S. Swearingen
1972—Aaron J. Teller
1973—C. J. King
1974—H. C. Hottel
1975—Alan S. Michaels
1976—Arthur E. Humphrey
1977—Arthur M. Squires
1978—Vern W. Weekman, Jr.
1979—James R. Fair
1980—John H. Seinfeld
1981—Bernard S. Lee
1982—Kenneth B. Bischoff

Founders Awards

1958—T. H. Chilton
    J. V. N. Dorr
    O. A. Hougen
    S. D. Kirkpatrick
    W. K. Lewis
1959—W. H. McAdams
    F. J. Curtis
1960—Walter G. Whitman
    Milton C. Whitaker
1961—George E. Holbrook
    Warren L. McCabe
1962—Barnett F. Dodge
    Joseph H. Koffolt
    J. Henry Rushton
1963—John J. Healy, Jr.
    Raymond P. Dinsmore
    Raymond P. Generaux
    Thomas K. Sherwood
1964—Donald L. Katz
    Mott Souders, Jr.
1965—Francis C. Frary
    Carl F. Prutton
1966—Manson Benedict
    Ernest W. Thiele
    Edward R. Weidlein
1967—William R. Collings
    Hoyt C. Hottel
    Chalmer G. Kirkbride

1968—William B. Franklin
    William N. Lacey
1970—Jerry McAfee
    William A. Cunningham
    Walter E. Lobo
    Donald B. Keyes
1971—Edwin R. Gilliland
    John J. McKetta, Jr.
    Paul D. V. Manning
1972—Joseph C. Elgin
    Donald A. Dahlstrom
1973—T. A. Burtis
    W. R. Marshall
    J. J. Martin
    R. L. Pigford
1974—W. H. Corcoran
    H. D. Guthrie
    M. S. Peters
1975—T. W. Tomkowit
    Theodore Weaver
1976—Lawrence K. Cecil
    Arthur L. Conn
    James R. Fair, Jr.
    Irving Liebson

(Founders Awards continued next page)

Founders Awards continued

1977—Kenneth E. Coulter
        Harold S. Kemp
        James G. Knudsen
        F. J. Van Antwerpen
1978—Ernest B. Christiansen
        Klaus D. Timmerhaus
1979—A. Sumner West

1980—Stuart W. Churchill
        Richard R. Hughes
        Sheldon Isakoff
1981—G. Burnet, Jr.
        J. C. Dart
        J. W. Davison
        J. Y. Oldshue
1982—H. W. Flood
        Ralph Landau

Warren K. Lewis Award

1963—Barnett F. Dodge
1964—Olaf A. Hougen
1965—Edwin F. Gilliland
1966—Richard H. Wilhelm
1967—Donald L. Katz
1968—Joseph H. Koffolt
1969—John J. McKetta, Jr.
1970—Robert L. Pigford
1971—Neal R. Amundson
1972—Thomas K. Sherwood

1973—W. L. McCabe
1974—R. Byron Bird
1975—Joseph C. Elgin
1976—Robert C. Reid
1977—Arthur B. Metzner
1978—Stuart W. Churchill
1979—Max S. Peters
1980—Carroll O. Bennett
1981—Rutherford Aris
1982—William H. Corcoran

Alpha Chi Sigma Award

1966—H. G. Drickamer
1967—Donald B. Broughton
1968—Klaus D. Timmerhaus
1969—Rutherford Aris
1970—A. E. Dukler
1971—John H. Sinfelt
1972—Charles D. Prater
1973—C. J. Pings

1974—Sheldon K. Friedlander
1975—Arnold A. Bondi
1976—Howard Brenner
1977—Eli Ruckenstein
1978—John A. Quinn
1979—Reuel Shinnar
1980—Roy Jackson
1981—W. E. Stewart
1982—Edward Wilson Merrill

R. H. Wilhelm Award

1973—Neal R. Amundson
1974—Michel Boudart
1975—Rutherford Aris
1976—James J. Carberry
1977—Joe Mauk Smith

1978—Paul B. Weisz
1979—Octave Levenspiel
1980—Charles N. Satterfield
1981—R. A. Schmitz
1982—Vern W. Weekman, Jr.

APPENDIX

Award for Service to Society

1973—L. K. Cecil
1974—Gerald A. Lessells
1975—John J. McKetta, Jr.
1976—Robert T. Jaske
1977—Frank W. Dittman

1978—Don E. Cox
1979—Gerald L. Decker
1980—Chris B. Earl
1981—R. H. Kummler
1982—Gary F. Bennett

Award in Chemical Engineering Practice

1974—Robert G. Heitz
1975—James R. Fair, Jr.
1976—Jacob M. Geist
1977—David Brown

1978—John W. Scott
1979—John Longwell
1980—David K. Beavon
1981—W. G. Schlinger

1982—Robert B. MacMullin

F. J. Van Antwerpen Award

1978—F. J. Van Antwerpen
1979—J. Henry Rushton

1980—George E. Holbrook
1981—Marx Isaacs

1982—W. Robert Marshall

EMINENT CHEMICAL ENGINEERS

Neal R. Amundson
Thomas Baron
Manson Benedict
R. Byron Bird
Theodore A. Burtis
Stuart W. Churchill
Donald A. Dahlstrom
W. Kenneth Davis
Thomas B. Drew
Harry G. Drickamer
James R. Fair
George E. Holbrook
Hoyt C. Hottel
Olaf A. Hougen
Arthur E. Humphrey

Donald L. Katz
Chalmer G. Kirkbride
Ralph Landau
W. Robert Marshall
Jerry McAfee
John J. McKetta, Jr.
Arthur B. Metzner
James Y. Oldshue
Max S. Peters
Robert L. Pigford
J. Henry Rushton
Klaus D. Timmerhaus
James Wei
James W. Westwater
Charles R. Wilke

INSTITUTE SECTIONS

Akron (Akron, Ohio) 1937
Alton-Wood River (Wood River, Ill.) 1955
Arizona (Phoenix, Ariz.) 1965
Atlanta (Atlanta, Ga.) 1952
Bartlesville (Bartlesville, Okla.) 1946
Baton Rouge (Baton Rouge, La.) 1943
Boston (Boston, Mass.) 1938
Central Alabama (Birmingham, Ala.) 1963
Central Arkansas (Little Rock, Ark.) 1972
Central Carolinas (Charlotte, N. C.) 1964
Central Florida (Lakeland, Fla.) 1958
Central Illinois (Tuscola, Ill.) 1958
Central Jersey (Princeton, N. J.) 1976
Central Ohio (Columbus, Ohio) 1937
Central Oklahoma (Oklahoma City, Okla.) 1953
Central Pennsylvania (Danville, Pa.) 1955
Central Virginia (Elkton, Va.) 1950
Charleston (Charleston, W. V.) 1940
Chattanooga (Chattanooga, Tenn.) 1968
Chicago (Chicago, Ill.) 1925
Cleveland (Cleveland, Ohio) 1944
Coastal Bend (Corpus Christi, Tex.) 1955
Coastal Carolinas (Wilmington, N. C.) 1976
Coastal Georgia (Savannah, Ga.) 1952
Columbia Pacific (Longview, Wash.) 1958
Columbia Valley (Richland, Wash.) 1950
Dallas (Dallas, Tex.) 1954
Dayton (Dayton, Ohio) 1966
Delaware Valley 1925
Delaware County Subsection 1932
Philadelphia Subsection 1932
Wilmington Subsection 1932
Detroit (Detroit, Mich.) 1945
East Tennessee (Kingsport, Tenn.) 1943
East Texas (Longview, Tex.) 1959
Eastern North Carolina (Raleigh, N. C.) 1962
El Dorado (El Dorado, Ark.) 1951
Fairfield County (Stamford, Conn.) 1960
Great Salt Lake (Salt Lake City, Utah) 1956
Guadalupe (Victoria, Tex.) 1968
Idaho (Idaho Falls, Id.) 1956
Indianapolis (Indianapolis, Ind.) 1958
Iowa (Cedar Rapids, Iowa) 1964
Joliet (Joliet, Ill.) 1974
Kansas City (Kansas City, Mo.) 1949
Knoxville-Oak Ridge (Oak Ridge, Tenn.) 1948

Lehigh Valley (Bethlehem, Pa.) 1960
Long Island (Farmingdale, N. Y.) 1961
Louisville (Louisville, Ky.) 1949
Maryland (Baltimore, Md.) 1935
Memphis (Memphis, Tenn.) 1956
Mid-Hudson (Beacon, N. Y.) 1960
Mid-Michigan (Midland, Mich.) 1952
Mobile (Mobile, Ala.) 1967
Mojave Desert (Trona, Calif.) 1957
Nashville (Nashville, Tenn.) 1952
National Capital (Washington, D.C.) 1950
Nebraska (Omaha, Nebr.) 1970
The Netherlands (The Hague) 1970
New Haven (New Haven, Conn.) 1955
New Jersey (Plainfield, N. J.) 1949
New Orleans (New Orleans, La.) 1932
New York (New York, N. Y.) 1933
North Alabama (Decatur, Ala.) 1942
North Jersey (Nutley, N. J.) 1958
Northeastern New York (Schenectady, N. Y.) 1953
Northern California (San Francisco, Calif.) 1946
Ohio Valley (Cincinnati, Ohio) 1946
Oregon (Salem, Oreg.) 1979
Palmetto (Columbia, S. C.) 1972
Peninsular Florida (Gainesville, Fla.) 1958
Pensacola (Pensacola, Fla.) 1956
Permian (Odessa, Tex.) 1965
Pittsburgh (Pittsburgh, Pa.) 1935
Puget Sound (Seattle, Wash.) 1947
Rhode Island (Providence, R. I.) 1953
Rio Grande (El Paso, Tex.) 1975
Rochester (Rochester, N. Y.) 1946
Rocky Mountain (Denver, Colo.) 1949
St. Louis (St. Louis, Mo.) 1933
San Diego (San Diego, Calif.) 1975
Savannah River (Aiken, S. C.) 1954
South Jersey (Moorestown, N. J.) 1972
South Texas (Houston, Tex.) 1942
Southeast Texas (Orange, Tex.) 1950
Southern California (Los Angeles, Calif.) 1946
Southwest Louisiana (Lake Charles, La.) 1958
Syracuse (Syracuse, N. Y.) 1957
Tappan Zee (Tarrytown, N. Y.) 1977
Terre Haute (Terre Haute, Ind.) 1951
Texas Gulf Coast (Brazosport, Tex.) 1974

APPENDIX

Texas Panhandle (Borger, Tex.) 1951
Tidewater Virginia (Richmond, Va.) 1953
Toledo (Toledo, Ohio) 1955
Tri-State (Ashland, Ky.) 1967
Tulsa (Tulsa, Okla.) 1948
Twin Cities (St. Paul, Minn.) 1948
Twin Tiers (Corning, N. Y.) 1982
Western Kentucky (Calvert City, Ky.) 1966
Western Massachusetts (Springfield, Mass.) 1947
Western New York (Buffalo, N. Y.) 1936
Western North Carolina (Hendersonville, N. C.) 1961
Western South Carolina (Clemson, S. C.) 1966
Wichita (Wichita, Ks.) 1970
Wisconsin (Milwaukee, Wisc.) 1960

ACCREDITED CURRICULA IN CHEMICAL ENGINEERING LEADING TO THE FIRST PROFESSIONAL DEGREE

Akron University of Akron OH (cooperative also) 1970
Alabama, University of, University, AL 1950
Arizona, University of, Tucson, AZ 1963
Arizona State University, Tempe, AZ 1966
Arkansas, University of, Fayetteville, AR 1952
Auburn University, Auburn, AL (cooperative also) 1950
Brigham Young University, Provo, UT 1961
Bucknell University, Lewisburg, PA 1942
California, University of, Berkeley, CA (cooperative also) 1952
California University of, Davis, CA 1966
California, University of, Santa Barbara, CA 1968
California Institute of Technology, Pasadena, CA 1926
California State Polytechnic University, Pomona, CA 1972
California State University, Long Beach, CA 1980
Carnegie-Mellon University, Pittsburgh, PA 1925
Case Western Reserve University, Cleveland, OH 1925
Catholic University of America, The Washington, DC 1967
Cincinnati, University of, Cincinnati, OH (cooperative also) 1925
Clarkson College of Technology, Potsdam, NY 1938
Clemson University, Clemson, SC 1959
Cleveland State University, Cleveland, OH (cooperative also) 1954-1963, 1966
Colorado, University of, Boulder, CO 1950
Colorado School of Mines, Golden, CO (Chemical and Pet. Ref. Eng.) 1956
Colorado State University, Ft. Collins, CO 1981
Columbia University, New York, NY 1925
Connecticut, University of, Storrs, CT 1964
Cooper Union, The, New York, NY 1941
Cornell University, Ithaca, NY 1936
Dayton, University of, Dayton, OH 1969
Delaware, University of, Newark, DE 1941
Detroit, University of, Detroit, MI (cooperative also) 1940-49, 1951
Drexel University, Philadelphia, PA (cooperative also) 1936-1947, 1949
Florida, University of, Gainesville, FL 1942
Georgia Institute of Technology, Atlanta, GA (cooperative also) 1938
Houston, University of, Houston, TX (cooperative also) 1957
Howard University, Washington, DC (cooperative also) 1977
Idaho, University of, Moscow, ID 1950
Illinois, University of, Urbana, IL (cooperative also) 1933
Illinois at Chicago Circle, University of, Chicago, IL (cooperative also) 1976
Illinois Institute of Technology, Chicago, IL (cooperative also) 1925
Iowa, University of, Iowa City, IA 1926-1950, 1953
Iowa State University, Ames, IA (cooperative also) 1925
Kansas, University of, Lawrence, KS 1949
Kansas State University, Manhattan, KS 1951

(continued next page)

Accredited Curricula continued

Kentucky, University of, Lexington, KY 1969
Lafayette College, Easton, PA 1956
Lamar University, Beaumont, TX (cooperative also) 1958
Lehigh University, Bethlehem, PA 1932-1951, 1953
Louisiana State University, Baton Rouge, LA 1939
Louisiana Technical University, Ruston, LA 1956
Louisville, University of, Louisville, KY (five-year program, master's degree, cooperative) 1935
Lowell, University of, Lowell, MA 1971
Maine, University of, Orono, ME 1950
Manhattan College, Riverdale, NY 1964
Maryland, University of, College Park, MD (cooperative also) 1942
Massachusetts, University of, Amherst, MA 1958
Massachusetts Institute of Technology, Cambridge, MA 1925
Michigan, University of, Ann Arbor, MI 1925
Michigan State University, East Lansing, MI 1954
Michigan Technological University, Houghton, MI 1947
Minnesota, University of, Minneapolis, MN 1925
Mississippi, University of, University, MS 1954
Mississippi State University, Mississippi State, MS (cooperative also) 1964
Missouri-Columbia, University of, Columbia, MO 1940
Missouri-Rolla, University of, Rolla, MO 1951
Montana State University, Bozeman, MT 1948
Nebraska-Lincoln, University of, Lincoln, NE 1954
New Hampshire, University of, Durham, NH 1964
New Jersey Institute of Technology, Newark, NJ (evening also) 1950
New Mexico, University of, Albuquerque, NM 1976
New Mexico State University, Las Cruces, NM 1967

New York, City College of the City University of, New York, NY 1953
New York, Polytechnic Institute of, Brooklyn, NY 1925
New York, State University of, Buffalo, NY 1966
North Carolina State University, Raleigh, NC (cooperative also) 1948
North Dakota, University of, Grand Forks, ND 1939-1942, 1951
Northeastern University, Boston, MA (cooperative also) 1942
Northwestern University, Evanston, IL (cooperative also) 1947
Notre Dame, University of, Notre Dame, IN 1949
Ohio State University, Columbus, OH 1925
Ohio University, Athens, OH 1963
Oklahoma, University of, Norman, OK 1940
Oklahoma State University, Stillwater, OK 1950-1976 (master's degree program) 1976
Oregon State University, Corvallis, OR 1942
Pennsylvania, University of, Philadelphia, PA 1936
Pennsylvania State University, University Park, PA 1936
Pittsburgh, University of, Pittsburgh, PA 1931-1965, 1967
Pratt Institute, Brooklyn, NY (cooperative also) 1939
Princeton University, Princeton, NJ 1934
Puerto Rico, University of, Mayaguez, PR (five-year program) 1970
Purdue University, West Lafayette, IN (cooperative also) 1933
Rensselaer Polytechnic Institute, Troy, NY (cooperative also) 1925
Rhode Island, University of, Kingston, RI 1954
Rice University, Houston, TX 1941
Rochester, University of, Rochester, NY 1941
Rose-Hulman Institute of Technology, Terre Haute, IN 1950
Rutgers, The State University, New Brunswick, NJ 1971
San Jose State University, San Jose, CA 1966

APPENDIX

South Carolina, University of, Columbia, SC 1956
South Dakota School of Mines and Technology, Rapid City, SD 1960
South Florida; University of, Tampa, FL 1976
Southern California, University of, Los Angeles, CA 1950-1952, 1960
Southwestern Louisiana, University of, Lafayette, LA 1956-1963, 1967
Stanford University, Stanford, CA 1959
Syracuse University, Syracuse, NY 1940
Tennessee, University of, Knoxville, TN (cooperative also) 1939
Tennessee Technological University, Cookeville, TN (cooperative also) 1970
Texas at Austin, University of, Austin, TX (cooperative also) 1943
Texas A & I University, Kingsville, TX 1978
Texas A & M University, College Station, TX (cooperative also) 1946
Texas Tech University, Lubbock, TX 1965
Toledo, University of, Toledo, OH 1964
Tri-State University, Angola, IN (cooperative also) 1979
Tufts University, Medford, MA 1952
Tulane University, New Orleans, LA 1954
Tulsa, University of, Tulsa, OK 1958
Utah, University of, Salt Lake City, UT 1952
Vanderbilt University, Nashville, TN 1949
Villanova University, Villanova, PA 1951
Virginia Polytechnic Institute and State University, Blacksburg, VA (cooperative also) 1938
Virginia, University of, Charlottesville, VA 1943
Washington, University of, Seattle, WA 1926
Washington State University, Pullman, WA 1951
Washington University, St. Louis, MO (cooperative also) 1948
Wayne State University, Detroit, MI 1950
West Virginia Institute of Technology, Montgomery, WV (cooperative also) 1972
West Virginia University, Morgantown, WV 1948
Wisconsin-Madison, University of, Madison, WI 1925
Worcester Polytechnic Institute, Worcester, MA 1942
Wyoming, University of, Laramie, WY 1974
Yale University, New Haven, CT 1982
Youngstown State University, Youngstown, OH 1974

CHRONOLOGICAL LIST OF MEETINGS

Meeting for Organization	Philadelphia, Pa.	June 22, 1908
1 annual	Pittsburgh, Pa.	Dec. 28-29, 1908
1 semiannual	Brooklyn, N.Y.	June 24-26, 1909
2 annual	Philadelphia, Pa	Dec. 3-11, 1909
2 semiannual	Niagara Falls, N.Y.	June 22-24, 1910
3 annual	New York, N.Y.	Dec. 7-10, 1910
3 semiannual	Chicago, Ill.	June 21-24, 1911
4 annual	Washington, D.C.	Dec. 20-23, 1911
4 semiannual (Inter. Congress)	New York, N.Y.	Sept. 5, 1912
5 annual	Detroit, Mich.	Dec. 4-7, 1912
5 semiannual	Boston, Mass.	June 25-28, 1913
6 annual	New York, N.Y.	Dec. 10-13, 1913
6 semiannual	Troy, N.Y.	June 17-20, 1914
7 annual	Philadelphia, Pa.	Dec. 2-4, 1914
7 semiannual	Los Angeles, Calif.	Aug. 16-18, 1915
7 semiannual	San Francisco, Calif.	Aug. 25-28, 1915
8 annual	Baltimore, Md.	Jan. 12-15, 1916
8 semiannual	Cleveland, Ohio	June 14-17, 1916

APPENDIX

Meetings continued

9 annual	New York, N.Y.	Jan 10-12, 1917
9 semiannual	Buffalo, N.Y.	June 20-22, 1917
10 annual	St. Louis, Mo.	Dec. 5-8, 1917
10 semiannual	Gorham and Berlin, N.H.	June 19-22, 1918
11 annual	Chicago, Ill.	Jan. 15-18, 1919
11 semiannual	Boston, Mass.	June 18-21, 1919
12 annual	Savannah, Ga.	Dec. 3-6, 1919
12 semiannual	Montreal, Canada	June 28-July 3, 1920
13 annual	New Orleans, La.	Dec. 6-9, 1920
13 semiannual	Detroit, Mich.	June 20-25, 1921
14 annual	Baltimore, Md.	Dec. 5-8, 1921
14 semiannual	Niagara Falls and Buffalo	June 19-22, 1922
15 annual	Richmond, Va.	Dec. 6-9, 1922
15 semiannual	Wilmington, Del.	June 20-23, 1923
16 annual	Washington, D.C.	Dec. 5-8, 1923
16 semiannual	Denver, Colo.	July 15-18, 1924
17 annual	Pittsburgh, Pa.	Dec. 3-6, 1924
17 semiannual	Providence, R.I.	June 23-17, 1925
Joint meeting with The Institution of Chemical Engineers	London, England	July 12-30, 1925
18 annual	Cincinnati, Ohio	Dec. 2-5, 1925
18 semiannual	Berlin, N.H.	June 21-23, 1926
19 annual	Atlanta-Birmingham	Dec. 6-10, 1926
19 semiannual	Cleveland, Ohio	May 31, June 1-3, 1927
20 annual	St. Louis, Mo.	Dec. 5-8, 1927
20 semiannual (with The Institution of Chemical Engineers)	Niagara Falls, N.Y. (Tour Eastern Canada and United States)	Aug. 18-Sept. 2, 1928
21 annual	New York, N.Y.	Dec. 4, 1928
21 semiannual	Philadelphia, Pa.	June 19-21, 1929
22 annual	Asheville, N.C.	Dec. 1-3, 1929
22 semiannual	Detroit, Mich.	June 4-6, 1930
23 annual	New Orleans, La.	Dec. 8-10, 1930
23 semiannual	Swampscott, Mass.	June 10-12, 1931
24 annual	Atlantic City, N.J.	Dec. 9-11, 1931
24 semiannual	Schenectady & Corning, N.Y.	June 15-17, 1932
25 annual	Washington, D.C.	Dec. 7-9, 1932
25 semiannual	Chicago, Ill.	June 14-16, 1933
26 annual	Roanoke, Va.	Dec. 12-14, 1933
26 semiannual	New York, N.Y.	May 14-16, 1934
27 annual	Pittsburgh, Pa.	Nov. 15-17, 1934
27 semiannual	Wilmington, Del.	May 13-15, 1935
28 annual	Columbus, Ohio	Nov. 13-15, 1935
28 semiannual	New York, N.Y.	June 12, 1936
29 annual	Baltimore, Md.	Nov. 11-13, 1936
29 semiannual	Toronto, Canada	May 26-28, 1937
30 annual	St. Louis, Mo	Nov. 17-19, 1937

APPENDIX

30 semiannual	White Sulphur Springs, W. Va.	May 9-11, 1938
31 annual	Philadelphia, Pa.	Nov. 9-11, 1938
31 semiannual	Akron, Ohio	May 15-17, 1939
32 annual	Providence, R.I.	Nov. 15-17, 1939
32 semiannual	Buffalo-Niagara Falls	May 13-15, 1940
33 annual	New Orleans, La.	Dec. 2-4, 1940
33 semiannual	Chicago, Ill.	May 19-21, 1941
34 annual	Virginia Beach, Va.	Nov. 3-5, 1941
34 semiannual	Boston, Mass.	May 11-13, 1942
35 annual	Cincinnati, Ohio	Nov. 16-17, 1942
35 semiannual	New York, N.Y.	May 10-11, 1943
36 annual	Pittsburgh, Pa.	Nov. 15-16, 1943
36 semiannual	Cleveland, Ohio	May 14-16, 1944
37 annual	St. Louis, Mo.	Nov. 19-21, 1944
38 annual	Chicago, Ill.	Dec. 16-19, 1945
1 national	New York, N.Y.	Feb. 27-28, 1946
2 national	Houston, Texas	March 31, Apr. 1-3, 1946
3 national	San Francisco, Calif.	Aug. 25-28, 1946
39 annual	Atlantic City, N.J.	Dec. 1-4, 1946
4 national	Louisville, Ky.	Feb. 16-19, 1947
5 national	St. Louis, Mo.	May 11-14, 1947
6 national	Buffalo, N.Y.	Sept. 29-Oct. 1, 1947
40 annual	Detroit, Mich.	Nov. 9-12, 1947
7 national	New Orleans, La.	Feb. 15-19, 1948
8 national	Cleveland, Ohio	May 9-12, 1948
9 national	French Lick, Ind.	Sept. 15-17, 1948
41 annual	New York, N.Y.	Nov. 7-10, 1948
10 national	Los Angeles, Calif.	March 6-9, 1949
11 national	Tulsa, Okla.	May 8-11, 1949
12 national	Montreal, Que., Canada	Sept. 6-8, 1949
42 annual	Pittsburgh, Pa.	Dec. 4-7, 1949
13 national	Houston, Texas	Feb. 26-March 1, 1950
14 national	Swampscott, Mass.	May 28-31, 1950
15 national	Minneapolis, Minn.	Sept. 10-13, 1950
43 annual	Columbus, Ohio	Dec. 3-6, 1950
16 national	White Sulphur Springs, W. Va.	March 11-14, 1951
17 national	Kansas City, Mo.	May 13-16, 1951
18 national	Rochester, N.Y.	Sept. 16-19, 1951
44 annual	Atlantic City, N.J.	Dec. 2-5, 1951
19 national	Atlanta, Ga.	March 16-19, 1952
20 national	French Lick, Ind.	May 11-14, 1952
21 national	Chicago, Ill.	Sept. 11-13, 1952
45 annual	Cleveland, Ohio	Dec. 7-10, 1952
22 national	Biloxi, Miss.	March 8-11, 1953
23 national (with the Chemical Institute of Canada)	Toronto, Can.	April 26-29, 1953

(continued next page)

Meetings continued

24 national	San Francisco, Calif.	Sept. 13-16, 1953
46 annual	St. Louis, Mo.	Dec. 13-16, 1953
25 national	Washington, D.C.	March 7-10, 1954
26 national	Springfield, Mass.	May 16-19, 1954
27 national Nuclear Engineering Congress	Ann Arbor, Michigan	June 20-25, 1954
28 national	Glenwood Springs, Colorado	Sept. 12-15, 1954
47 annual	New York, New York	Dec. 12-15, 1954
29 national	Louisville, Ky.	March 20-23, 1955
30 national	Houston, Texas	May 1-4, 1955
31 national	Lake Placid, N.Y.	Sept. 25-28, 1955
48 annual	Detroit, Mich.	Nov. 27-30, 1955
32 national	Los Angeles, Calif.	Feb. 26-29, 1956
33 national	New Orleans, La.	May 6-9, 1956
34 national	Pittsburgh, Pa.	Sept. 9-12, 1956
49 annual	Boston, Mass.	Dec. 9-12, 1956
35 national	White Sulphur Springs, W. Va.	March 3-6, 1957
36 national	Seattle, Wash.	June 9-12, 1957
37 national	Baltimore, Md.	Sept. 15-18, 1957
50 annual	Chicago, Ill.	Dec. 8-11, 1957
With Chemical Institute of Canada	Montreal, Canada	April 20-23, 1958
Golden Jubilee	Philadelphia, Pa.	June 22-27, 1958
38 national	Salt Lake City, Utah	Sept. 21-24, 1958
51 annual	Cincinnati, Ohio	Dec. 7-10, 1958
39 National	Atlantic City, N.J.	March 15-18, 1959
40 National	Kansas City, Mo.	May 17-20, 1959
41 National	St. Paul, Minn.	Sept. 27-30, 1959
52 Annual	San Francisco, Calif.	Dec. 6-9, 1959
42 National	Atlantic, Ga.	Feb. 21-24, 1960
With the Instituto Mexicano de Ingenieros Quimicos	Mexico City	June 19-20, 1960
43 National	Tulsa, Okla.	Sept. 25-28, 1960
53 Annual	Washington, D.C.	Dec. 4-7, 1960
44 National (with First Petro- chemical & Refining Exposition)	New Orleans, La.	Feb. 26-Mar. 1, 1961
With the Chemical Institute of Canada	Cleveland, Ohio	May 7-10, 1961
45 National	Lake Placid, N.Y.	Sept. 24-27, 1961
54 Annual	New York, N.Y.	Dec. 3-7, 1961
46 National	Los Angeles, California	Feb. 4-7, 1962
47 National	Baltimore, Maryland	May 20-23, 1962
48 National	Denver, Colorado	Aug. 26-29, 1962
55 Annual	Chicago, Illinois	Dec. 2-6, 1962

APPENDIX

49 National (with Second Petrochemical & Refining Exposition)	New Orleans, Louisiana	March 10-14, 1963
50 National	Buffalo, New York	May 5-8, 1963
51 National	Puerto Rico	Sept. 29-Oct. 2, 1963
56 Annual	Houston, Texas	Dec. 1-5, 1963
52 National	Memphis, Tennessee	Feb. 2-5, 1964
53 National	Pittsburgh, Pennsylvania	May 17-20, 1964
54 National	Las Vegas, Nevada	Sept. 20-23, 1964
57 Annual	Boston, Mass.	Dec. 6-10, 1964
55 National (with Third Petrochemical & Refining Exposition)	Houston, Texas	Feb. 7-11, 1965
56 National	San Francisco, Calif.	May 16-19, 1965
With The Institution of Chemical Engineers, London	London, England	June 13-17, 1965
57 National	Minneapolis, Minnesota	Sept. 26-29, 1965
58 Annual	Philadelphia, Pennsylvania	Dec. 5-9, 1965
58 National	Dallas, Texas	Feb. 6-9, 1966
59 National	Columbus, Ohio	May 15-18, 1966
60 National	Atlantic City, N.J.	Sept. 18-21, 1966
59 Annual	Detroit, Michigan	Dec. 4-8, 1966
61 National (with Fourth Petrochemical & Refining Exposition)	Houston, Texas	Feb. 19-23, 1967
62 National	Salt Lake City, Utah	May 21-24, 1967
With the Instituto Mexicano de Ingenieros Quimicos	Mexico City	Sept. 24-27, 1967
60 Annual	New York, New York	Nov. 26-30, 1967
63 National	St. Louis, Mo.	Feb. 18-21, 1968
Materials Engineering & Sciences Conference and Exposition	Philadelphia, Pa.	March 31-April 4, 1968
With Instituto Puerto Ricano de Ingenieros Quimicos	Tampa, Fla.	May 19-22, 1968
With Canadian Society of Chemical Engineers and Institution of Chemical Engineers, London	Montreal, Canada	Sept. 22-25, 1968
61 Annual	Los Angeles, Calif.	Dec. 1-5, 1968

(continued next page)

APPENDIX

Meetings continued

64 National (with Fifth Petrochemical & Refining Exposition	New Orleans, La.	March 16-20, 1969
65 National	Cleveland, Ohio	May 4-7, 1969
66 National	Portland, Oregon	Aug. 24-27, 1969
62 Annual	Washington, D.C.	Nov. 16-20, 1969
67 National	Atlanta, Ga.	Feb. 16-19, 1970
With Instituto Puerto Ricano de Ingenieros Quimicos	San Juan, P.R.	May 17-20, 1970
With the Instituto Mexicano de Ingeniero Quimicos	Denver, Colo.	Aug. 31-Sept. 3, 1970
63 Annual	Chicago, Ill.	Nov. 29-Dec. 3, 1970
68 National & Sixth Petrochemical & Petroleum Refining Exposition	Houston, Texas	Feb. 28-March 4, 1971
69 National	Cincinnati, Ohio	May 16-19, 1971
70 National	Atlantic City, N.J.	Aug. 29-Sept. 1, 1971
64 Annual	San Francisco, Calif.	Nov. 28-Dec. 2, 1971
71 National	Dallas, Texas	Feb. 20-23, 1972
72 National	St. Louis, Mo.	May 21-24, 1972
73 National	Minneapolis, Minn.	Aug. 27-30, 1972
1 Pacific Chemical Engineering Conference	Kyoto, Japan	Oct. 11-14, 1972
65 Annual	New York, New York	Nov. 26-30, 1972
74 National & Seventh Petrochemical & Petrochemical & Petroleum Refining Exposition	New Orleans, La.	March 11-15, 1973
75 National	Detroit, Michigan	June 3-6, 1973
4th Joint AIChE-CSChE.	Vancouver, Canada	Sept. 9-12, 1973
66 Annual	Philadelphia, Pa.	Nov. 11-15, 1973
76 National	Tulsa, Oklahoma	March 10-13, 1974
77 National	Pittsburgh, Pa.	June 2-5, 1974
78 National	Salt Lake City, Utah	Aug. 18-21, 1974
Joint AIChE-VDI Meeting	Munich, Germany	Sept. 11-14, 1974
67 Annual	Washington, D.C.	Dec. 1-5, 1974
79 National & Eighth Petrochemical & Refining Exposition	Houston, TX	March 16-20, 1975
6 Inter-American Congress of Chemical Engineering	Caracas, Venezuela	July 13-16, 1975
80 National	Boston, MA	Sept. 7-10, 1975

APPENDIX

68 Annual	Los Angeles, CA	Nov. 16-20, 1975
81 National	Kansas City, MO	April 11-14, 1976
World Congress on Chemical Engineering	Amsterdam, The Netherlands	June 28-July 1, 1976
82 National	Atlantic City, NJ	Aug. 29-Sept. 1, 1976
69 Annual	Chicago, IL	Nov. 29-Dec. 2, 1976
83 National & Ninth Petrochemical & Refining Exposition	Houston, TX	March 20-24, 1977
2 PACHEC & 7 Interamerican Congress of Chemical Engineering	Denver, CO	Aug. 28-31, 1977
70 Annual	New York, NY	Nov. 13-17, 1977
84 National	Atlanta, GA	Feb. 26-March 1, 1978
85 National & First Chemical Plant Equipment Exposition	Philadelphia, PA	June 4-8, 1978
71 Annual	Miami, FL	Nov. 12-16, 1978
86 National & Tenth Petrochemical & Refining Exposition	Houston, TX	April 1-5, 1979
8th Inter-American Congress of Chemical Engineering	Bogota, Columbia	Aug. 5-12, 1979
87 National	Boston, MA	Aug. 19-22, 1979
72 Annual	San Francisco, CA	Nov. 25-29, 1979
88 National & 2nd Chemical Plant Equipment Exposition	Philadelphia, PA	June 8-12, 1980
89 National	Portland, OR	Aug. 17-20, 1980
With Instituto Mexicano de Ingenieros Quimicos	Acapulco, Mexico	Oct. 15-17, 1980
73 Annual Meeting	Chicago, IL	Nov. 16-20, 1980
1981 Spring National & Eleventh Petrochemical & Refining Exposition	Houston, TX	April 5-9, 1981
1981 Summer National Meeting	Detroit, MI	Aug. 16-19, 1981
2nd World Congress of Chemical Engineering	Montreal, Canada	Oct. 4-7, 1981
1981 Annual Meeting	New Orleans, LA	Nov. 8-12, 1981
1982 Winter National Meeting	Orlando, FL	Feb. 28-March 3, 1982
1982 Spring National Meeting & Third Chemical Plant Equipment Exposition	Anaheim, CA	June 6-10, 1982
1982 Summer National Meeting	Cleveland, OH	Aug. 29-Sept. 1, 1982
IChEME Double Jubilee	London, England	Oct. 13-15, 1982
1982 Annual Meeting	Los Angeles, CA	Nov. 14-19, 1982

(continued next page)

Meetings continued

1983 Spring National Meeting & Twelfth Petrochemical & Refining Exposition	Houston, TX	March 27-31, 1983
PACHEC III	Seoul, Korea	May 5-11, 1983
1983 Summer National Meeting	Denver, CO	Aug. 28-31, 1983
1983 Annual Meeting & Diamond Jubilee	Washington, D.C.	Oct. 30-Nov. 4, 1983

MEMBERSHIP STATISTICS

YEAR	TOTAL	YEAR	TOTAL
1908	40	1945	4,860
1909	101	1946	5,769
1910	118	1947	6,727
1911	145	1948	7,914
1912	170	1949	8,938
1913	195	1950	9,677
1914	214	1951	10,411
1915	220	1952	11,428
1916	230	1953	12,588
1917	270	1954	13,600
1918	283	1955	14,453
1919	306	1956	15,082
1920	344	1957	15,870
1921	454	1958	17,973
1922	529	1959	18,657
1923	565	1960	19,709
1924	613	1961	21,420
1925	644	1962	23,257
1926	681	1963	24,272
1927	723	1964	25,900
1928	788	1965	27,322
1929	805	1966	29,070
1930	872	1967	31,466
1931	944	1968	34,103
1932	1,017	1969	36,555
1933	1,066	1970	38,224
1934	1,099	1971	38,622
1935	1,157	1972	39,057
1936	1,321	1973	39,030
1937	1,435	1974	39,263
1938	1,733	1975	38,968
1939	1,984	1976	39,843
1940	2,255	1977	40,011
1941	2,527	1978	42,241
1942	2,923	1979	45,248
1943	3,347	1980	47,120
1944	3,929	1981	48,950
	1982		51,445